A Brief Introduction to Matrices and Vectors

Preliminary Edition

A Brief Introduction to Matrices and Vectors

Preliminary Edition

Jimmy T. Arnold

Lee W. Johnson

R. Dean Riess

Virginia Polytechnic Institute and State University

 **ADDISON-WESLEY**

An imprint of Addison Wesley Longman, Inc.

Reading, Massachusetts • Menlo Park, California • New York • Harlow, England
Don Mills, Ontario • Sydney • Mexico City • Madrid • Amsterdam

Reproduced by Addison-Wesley Publishing Company Inc. from camera-ready copy supplied by the authors.

Copyright © 1998 Addison Wesley Longman.

ISBN   0-201-30828-2

1 2 3 4 5 6 7 8 9 10 VG 00999897

Preface

This text is designed for first-year college students and advanced high school students. It can be used by itself or in conjunction with Internet-based materials that have been developed for the text.

The text centers on:

- Solving systems of linear equations
- Carrying out basic matrix operations
- Working with geometric vectors as they are used in physics and engineering

Our presentation emphasizes geometric intuition and practical computation, things students increasingly need in today's computer-intensive science and engineering courses.

Our text is organized into short 3 - 4 page sections, with one major topic per section. This approach allows students to keep their focus. In addition, short sections mean that the exercises provide immediate feedback.

Clarity is one of our primary goals. Because brevity is an aid to clarity, we try to get to the point quickly, with a minimum of the bells and whistles that clutter so many contemporary textbooks.

Contents

Chapter 4

A Brief Introduction to Matrices and Vectors

Preliminary Edition

1.1 INTRODUCTION

In mathematics, science, engineering, economics, and the social sciences, the need to solve systems of linear equations arises frequently and in a variety of settings. We give three specific examples below.

Three examples of linear systems

As our first example, consider the system of linear equations

(1)
$$
\begin{aligned}
I_1 - I_2 - I_3 &= 0 \\
20I_1 + 10I_2 \quad &= 20 \\
-10I_2 + 10I_3 &= 10.
\end{aligned}
$$

The variables I_1, I_2, and I_3 represent the unknown currents (in amperes) in the electrical circuit shown in Figure 1.

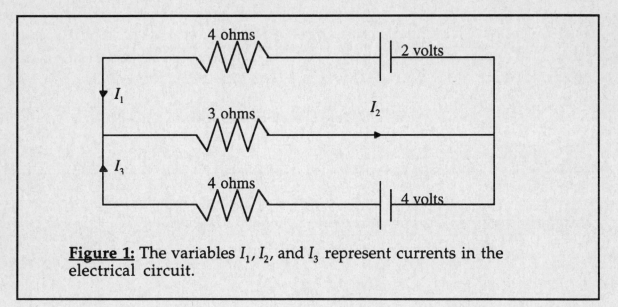

Figure 1: The variables I_1, I_2, and I_3 represent currents in the electrical circuit.

As a second illustration of a system of linear equations, suppose an island is divided into three regions A, B, and C. Moreover, suppose the yearly migration of a certain species among these three regions is described by the following table:

	From A	From B	From C
To A	70%	15%	10%
To B	15%	80%	30%
To C	15%	5%	60%

Table 1 Yearly migration rates from one region to another

Now, suppose we take a census and find that the current population consists of 300 individuals in region A, 350 in region B, and 200 in region C. If we want to know what the population distribution was last year, we can find it by solving the following system of equations:

$$
\begin{aligned}
.70x_1 + .15x_2 + .10x_3 &= 300 \\
.15x_1 + .80x_2 + .30x_3 &= 350 \\
.15x_1 + .05x_2 + .60x_3 &= 200.
\end{aligned}
$$

(2)

In the system above, x_1 represents the number of individuals in region A, x_2 the number in region B, and x_3 the number in region C.

As a final example, consider the simple data-fitting problem of passing a quadratic polynomial through three points in the xy-plane. In particular, what are values a, b, and c such that the graph of the quadratic polynomial $p(x) = ax^2 + bx + c$ passes through the three points $(-2, 21)$, $(1, 6)$, $(2, 13)$? To find these coefficents, we can solve the system of equations

$$
\begin{aligned}
4a - 2b + c &= 21 \\
a + b + c &= 6 \\
4a + 2b + c &= 13.
\end{aligned}
$$

(3)

Recognizing linear equations

The equation

$$-3x_1 + 2x_2 + x_3 = 6$$

is a *linear equation*. Another example of a linear equation is

$$3v - 4w = 1 + 2x - 7y.$$

On the other hand, the following are *nonlinear equations*:

$$3\sqrt{x} + 2xy = 4$$

and

$$y + 2 \ln x = 6.$$

In general, a *linear equation with n unknowns* is an equation that can be placed in the form

$$a_1 x_1 + a_2 x_2 + \cdots + a_n x_n = b.$$

In the equation above, $a_1, a_2, \ldots, a_n$, and b denote known constants while $x_1, x_2, \ldots, x_n$ are the *variables* or *unknowns*.

The key characteristic of a linear equation is that the variables appear to the first power only. In particular, a linear equation never contains products of variables (such as $x_1 x_2$), functions of variables (such as $\sin x_2$), or powers of variables such as $x_3{}^5$.

Example 1   Which of the following equations are linear equations?

(a) $x_1 + 2x_1 x_2 + 3x_3 = 4$ (b) $x_1{}^{-1} + 3 x_2 = 9$

(c) $\tan x_1 + 2x_2 = 0$ (d) $3x_1 - x_2 = x_3 + 1$

Solution: Only equation **(d)** is linear. The terms $x_1 x_2$, $x_1{}^{-1}$, and $\tan x_1$ are all nonlinear.   ■

Solutions to systems of equations

Consider the linear equation

$$a_1 x_1 + a_2 x_2 + \cdots + a_n x_n = b.$$

A _solution_ to this equation is any set of numbers $s_1, s_2, \ldots, s_n$ such that the equation is satisfied.

As a simple example, let us consider the equation

$$-3x_1 + 2x_2 + x_3 = 6.$$

The numbers $x_1 = 1$, $x_2 = 3$, $x_3 = 3$ give one solution to this equation as we see below:

$$-3\,(1) + 2\,(3) + (3) = 6.$$

Similarly, a solution to a _system_ of equations is any set of numbers $s_1, s_2, \ldots, s_n$ that simultaneously satisfies every equation in the system. For example, consider the following system of two linear equations in two unknowns:

$$x_1 - x_2 = 5$$
$$2x_1 - x_2 = 7$$

As you can easily verify, the choice $x_1 = 2$ and $x_2 = -3$ satisfies both of the equations and hence is a solution to the system.

In a like manner, you can verify that system (1) has solution $I_1 = 1$, $I_2 = 0$, and $I_3 = 1$. This solution therefore determines the currents in the electrical network shown in Figure 1.

The population model described by system (2) has solution $x_1 = 13700/41 \approx 334$, $x_2 = 11900 \approx 290$, $x_3 = 9250/41 \approx 225$. Thus, the model estimates that one year ago there were 334, 290, and 225 individuals in regions A, B, and C respectively.

Finally, we see that system (3) has a solution given by $a = 3$, $b = -2$, $c = 5$. Therefore, $p(x) = 3x^2 - 2x + 5$ is a quadratic polynomial whose graph passes through the three points $(-2, 21)$, $(1, 6)$, and $(2, 13)$.

Consistent and inconsistent linear systems

We will frequently refer to a system of m linear equations in n unknowns as an $(m \times n)$ linear system. If $m = n$ (so that there are exactly as many equations as there are unknowns) we will call the system a _square_ system. If a system is not square, it is called a _rectangular_ system.

If an $(m \times n)$ system has at least one solution, then we say the system is _consistent_. If the system has no solutions whatsoever, then it is called _inconsistent_. For example, the following square system is obviously inconsistent:

$$x_1 + x_2 = 0$$
$$x_1 + x_2 = 1$$

Augmented matrices

All the essential information about a system of linear equations can be read off from the coefficients of the system. So, when we want to describe a system, it is customary to use a kind of shorthand based on the coefficients.

As an illustration, consider the system below

$$2x_1 + 3x_2 - x_3 = 4$$
$$-x_1 + x_2 - 4x_3 = 2$$
$$3x_1 - x_2 + 2x_3 = 1$$

If we suppress the variables and the equal signs, we can represent this system in terms of its _augmented matrix_

$$\begin{bmatrix} 2 & 3 & -1 & 4 \\ -1 & 1 & -4 & 2 \\ 3 & -1 & 2 & 1 \end{bmatrix}.$$

Note that the augmented matrix contains all the information necessary to reconstruct the original system—we only have to "decode" the augmented matrix.

Example 2 Each matrix is the augmented matrix for a system of linear equations. Give the system.

(a) $\begin{bmatrix} 1 & 3 & 1 \\ 2 & 1 & 3 \end{bmatrix}$

(b) $\begin{bmatrix} 3 & 2 & 5 & -7 \\ 1 & 4 & -1 & 6 \\ 2 & 1 & 3 & -5 \end{bmatrix}$

(c) $\begin{bmatrix} 2 & 1 & 4 \\ 1 & -1 & 3 \\ -2 & 3 & 1 \end{bmatrix}$

(d) $\begin{bmatrix} 1 & -1 & 1 & 3 \\ 1 & 6 & 2 & 4 \end{bmatrix}$

Solution Decoding the various augmented matrices, we find the systems:

(a)
$$x_1 + 3x_2 = 1$$
$$2x_1 + x_2 = 3$$

(b)
$$3x_1 + 2x_2 + 5x_3 = -7$$
$$x_1 + 4x_2 - x_3 = 6$$
$$2x_1 + x_2 + 3x_3 = -5$$

(c)
$$2x_1 + x_2 = 4$$
$$x_1 - x_2 = 3$$
$$-2x_1 + 3x_2 = 1$$

(d)
$$x_1 - x_2 + x_3 = 3$$
$$x_1 + 6x_2 + 2x_3 = 4$$

■

Exercises 1.1

Which of the equations in Exercises 1-4 are linear?

1. $x_1 + 2x_3 = 3$

2. $x_1 x_2 + x_2 = 1$

3. $x_1 - x_2 = \sin^2 x_1 + \cos^2 x_2$

4. $\pi x_1 + \sqrt{7} x_2 = \sqrt{3}$

5. Check each of (a)-(d) to determine if it is a solution for the system of equations

$$
\begin{array}{rcrcrcr}
x_1 & + & 2x_2 & + & 4x_3 & = & 8 \\
x_1 & + & 3x_2 & + & 7x_3 & = & 10 \\
2x_1 & + & 3x_2 & + & 5x_3 & = & 14
\end{array}
$$

(a) $x_1 = 2$, $x_2 = 1$, $x_3 = 1$ (b) $x_1 = 4$, $x_2 = 2$, $x_3 = 0$

(c) $x_1 = 6$, $x_2 = -1$, $x_3 = 1$ (d) $x_1 = 10$, $x_2 = 1$, $x_3 = -1$

6. Check each of (a)-(d) to determine if it is a solution for the system of equations

$$
\begin{array}{rcrcrcrcr}
x_1 & - & 2x_2 & - & 7x_3 & & & = & 0 \\
x_1 & - & x_2 & - & 4x_3 & + & x_4 & = & 0
\end{array}
$$

(a) $x_1 = 0$, $x_2 = 0$, $x_3 = 0$, $x_4 = 0$ (b) $x_1 = 1$, $x_2 = -3$, $x_3 = 1$, $x_4 = 0$

(c) $x_1 = -2$, $x_2 = -1$, $x_3 = 0$, $x_4 = 1$ (d) $x_1 = 3$, $x_2 = -2$, $x_3 = 1$, $x_4 = -1$

In Exercises 7 and 8, show that the given system of equations is inconsistent.

7.
$$
\begin{aligned}
x_1 - 2x_2 &= 1 \\
2x_1 - 4x_2 &= -1
\end{aligned}
$$

8.
$$
\begin{aligned}
x_1 + x_2 &= 1 \\
x_2 + x_3 &= 1 \\
x_1 + 2x_2 + x_3 &= 0
\end{aligned}
$$

In Exercises 9-12 you are given the augmented matrix for a system of equations. Display the corresponding system of equations.

9.
$$
\begin{bmatrix}
1 & -1 & 3 & 4 \\
2 & 0 & 1 & 6 \\
0 & -1 & 3 & 1
\end{bmatrix}
$$

10.
$$
\begin{bmatrix}
3 & -1 & 0 \\
1 & 2 & 0 \\
-2 & 1 & 0
\end{bmatrix}
$$

11.
$$
\begin{bmatrix}
1 & 0 & 0 & 3 \\
0 & 1 & 0 & -1 \\
0 & 0 & 1 & 2
\end{bmatrix}
$$

12.
$$
\begin{bmatrix}
1 & 0 & 2 & -5 & 3 \\
0 & 1 & -3 & 2 & 1
\end{bmatrix}
$$

In Exercises 13-16 display the augmented matrix for the given system.

13.
$$
\begin{aligned}
x_1 - x_2 &= -1 \\
x_1 + x_2 &= 3
\end{aligned}
$$

14.
$$
\begin{aligned}
x_1 + x_2 - x_3 &= 2 \\
2x_1 - x_3 &= 1
\end{aligned}
$$

15.
$$
\begin{aligned}
3x_1 - 2x_2 - 4 &= 0 \\
x_1 + x_2 + 1 &= 0
\end{aligned}
$$

16.
$$
\begin{aligned}
x_1 - 2x_3 &= 5 - 3x_2 + x_4 \\
x_3 &= -1 - 7x_4
\end{aligned}
$$

1.2 GAUSS-JORDAN ELIMINATION

This section introduces Gauss-Jordan elimination, a practical and systematic method for solving systems of linear equations.

In order to keep the presentation as free from complications as possible, we will restrict ourselves in this section to square systems that have unique solutions. Later, in Sections 3 and 4, we will treat rectangular systems and systems that have no solutions or infinitely many solutions.

The solution procedure we want to discuss, Gauss-Jordan elimination, is based on a idea that is probably familiar to you—that of using elementary operations to eliminate some of the variables from an equation.

Elementary operations

We can introduce elementary operations with an example. Let us consider how we might go about solving the two linear systems:

$$\textbf{(a)} \quad \begin{aligned} x_1 - x_2 &= 1 \\ -x_2 &= -1 \end{aligned} \qquad\qquad \textbf{(b)} \quad \begin{aligned} x_1 - x_2 &= 1 \\ 2x_1 - 3x_2 &= 1 \end{aligned}$$

Because system **(a)** has a "triangular" form, it is not difficult to solve. In particular, multiplying the second equation by -1 tells us that $x_2 = 1$. Substituting $x_2 = 1$ in the first equation and solving for x_1 yields $x_1 = 2$.

It is a little harder to solve system **(b)**. But, examining the system, we see that we can eliminate the variable x_1 in the second equation if we add -2 times the first equation to the second equation (a shorthand representation for this operation is $E_2 - 2E_1$).

The operation $E_2 - 2E_1$ transforms system **(b)** into system **(a)**, and we just saw how to solve system **(a)**. Therefore, we know system **(b)** has the same solution, namely $x_1 = 2, x_2 = 1$. Our method for solving system **(b)** reminds us of an important mathematical principle: when we are confronted with a difficult problem to solve, we should try to reduce it to a simpler problem that has the same solution.

In terms of solving linear systems, there are three operations which can be used to simplify the system without changing the solution. These operations are called _**elementary operations**_ and they are:

1. Interchange any two equations

2. Multiply both sides of any equation by a nonzero constant

3. Replace any equation by adding a constant multiple of any other equation to it

The following example illustrates how we can use elementary operations to eliminate variables and thereby obtain a simpler system.

Example 1   Use elementary operations to eliminate the variable x_1 from the second and third equations

$$x_1 + 2x_2 + 2x_3 = 3$$
$$2x_1 + 7x_2 + 6x_3 = 7$$
$$x_1 + x_2 - 4x_3 = 6$$

Solution: If we multiply equation 1 by -2 and add the result to equation 2, we will eliminate the variable x_1 from the second equation. Carrying out the elementary operation $E_2 - 2E_1$, we obtain

$$x_1 + 2x_2 + 2x_3 = 3$$
$$3x_2 + 2x_3 = 1$$
$$x_1 + x_2 - 4x_3 = 6$$

Similarly, to eliminate x_1 from equation 3, we can multiply equation 1 by -1 and add the result to the third equation. Performing the operation $E_3 - E_1$ leads us to

$$x_1 + 2x_2 + 2x_3 = 3$$
$$3x_2 + 2x_3 = 1$$
$$-x_2 - 6x_3 = 3$$

■

Note, in Example 1, that we have essentially replaced the problem of solving a system of three equations by the simpler one of solving a system of two equations. In particular, once we solve the last two equations for x_2 and x_3, we can insert those values in the first equation and solve for x_1.

It can be shown that if we use any of the three elementary operations to create a new system, then the new system has exactly the same solutions as the original system. We call two systems having the same solution set _equivalent_ systems. Thus, whenever we apply an elementary operation, we produce a system equivalent to the one we started with.

Elementary row operations

When we use augmented matrices to represent linear systems, the above elementary operations are done on the rows of the augmented matrix instead of on the equations themselves. In this case, the operations are called _elementary row operations._ The notation we use to describe these elementary row operations is as follows:

Notation	Elementary row operation
$R_i \leftrightarrow R_j$	Interchange row i and row j
kR_i	Multiply row i by the nonzero constant k
$R_i + kR_j$	k times row j is added to row i

In Example 1 we illustrated how to use elementary operations to simplify a system of equations. In a like manner, we can use elementary row operations to simplify the augmented matrix for a system. Furthermore, if we systematically eliminate variables in a left-to-right fashion, we can avoid the possibility that the work done in one step of the process will undo the work we did at a previous step. This systematic elimination of variables in a left-to-right fashion is called *Gauss-Jordan elimination*.

We first illustrate Gauss-Jordan elimination in an example and then describe it more thoroughly.

Example 2 Solve the system

$$x_1 - 2x_2 = -5$$
$$-2x_1 + x_2 = 7$$

Solution: To eliminate x_1 in the second equation, we add twice the first equation to the second:

$$x_1 - 2x_2 = -5$$
$$- 3x_2 = -3$$

We now simplify the second equation by dividing both sides by -3:

$$x_1 - 2x_2 = -5$$
$$x_2 = 1$$

To eliminate x_2 in the first equation, we add two times the second equation to the first:

$$x_1 \qquad = -3$$
$$x_2 = 1$$

This final system is as simple as it can possibly be. We can read off the solution, namely $x_1 = -3$ and $x_2 = 1$. ∎

In terms of the shortcuts provided by using augmented matrices, the solution process of Example 2 can be carried out as follows:

$$\begin{bmatrix} 1 & -2 & -5 \\ -2 & 1 & 7 \end{bmatrix} \xrightarrow{R_2 + 2R_1} \begin{bmatrix} 1 & -2 & -5 \\ 0 & -3 & -3 \end{bmatrix} \xrightarrow{(-1/3)R_2} \begin{bmatrix} 1 & -2 & -5 \\ 0 & 1 & 1 \end{bmatrix}$$

$$\xrightarrow{R_1 + 2R_2} \begin{bmatrix} 1 & 0 & -3 \\ 0 & 1 & 1 \end{bmatrix}$$

Because of the simplicity provided by augmented matrices, we shall use them throughout the text whenever we need to solve a linear system. Finally, we shall call two matrices _**row equivalent**_ whenever one can be obtained from the other by a series of elementary row operations.

Gauss-Jordan elimination

When applied to the augmented matrix, Gauss-Jordan elimination consists of a series of "normalize–eliminate" steps. That is, we begin by multiplying the first row of the matrix by a nonzero constant in order to make the $(1, 1)$ entry equal to 1; this is a normalization step. We next add multiples of row 1 to the other rows in order to introduce zeros below the 1 in the $(1, 1)$ spot; these are elimination steps. We now move to the second column, normalize the $(2, 2)$ entry, and then add multiples of row 2 to the other rows in order to introduce zeros above and below the 1 in the $(2, 2)$ spot. Moving column by column, we continue these "normalize–eliminate" steps until we have swept through the entire matrix. (During the process, it may sometimes be necessary to perform a row interchange in order to get a nonzero entry in the (k, k) entry of the matrix.)

The example below illustrates Gauss-Jordan elimination for a linear system of three equations in three unknown.

Example 3   Solve the system

$$\begin{aligned} x_1 + 2x_2 - x_3 &= 5 \\ x_1 + x_2 + x_3 &= 2 \\ 2x_1 + 3x_2 - x_3 &= 8 \end{aligned}$$

Solution: Using augmented matrices, the solution proceeds as follows:

$$\begin{bmatrix} 1 & 2 & -1 & 5 \\ 1 & 1 & 1 & 2 \\ 2 & 3 & -1 & 8 \end{bmatrix} \xrightarrow[\begin{subarray}{c} R_2-R_1 \\ R_3-2R_1 \end{subarray}]{} \begin{bmatrix} 1 & 2 & -1 & 5 \\ 0 & -1 & 2 & -3 \\ 0 & -1 & 1 & -2 \end{bmatrix} \xrightarrow{-R_2} \begin{bmatrix} 1 & 2 & -1 & 5 \\ 0 & 1 & -2 & 3 \\ 0 & -1 & 1 & -2 \end{bmatrix}$$

$$\xrightarrow[\begin{subarray}{c} R_1-2R_2 \\ R_3+R_2 \end{subarray}]{} \begin{bmatrix} 1 & 0 & 3 & -1 \\ 0 & 1 & -2 & 3 \\ 0 & 0 & -1 & 1 \end{bmatrix} \xrightarrow{-R_3} \begin{bmatrix} 1 & 0 & 3 & -1 \\ 0 & 1 & -2 & 3 \\ 0 & 0 & 1 & -1 \end{bmatrix}$$

$$\xrightarrow[\begin{subarray}{c} R_1-3R_3 \\ R_2+2R_3 \end{subarray}]{} \begin{bmatrix} 1 & 0 & 0 & 2 \\ 0 & 1 & 0 & 1 \\ 0 & 0 & 1 & -1 \end{bmatrix}$$

The final matrix above is the augmented matrix for the simple system

$$\begin{aligned} x_1 \quad\quad\quad &= 2 \\ x_2 \quad\quad &= 1 \\ x_3 &= -1. \end{aligned}$$

Hence the solution to the original system is $x_1 = 2$, $x_2 = 1$, $x_3 = -1$. ∎

Exercises 1.2

In Exercises 1-6, use elementary operations to obtain an equivalent system of equations where only the first equation contains the variable x_1 and in the first equation x_1 has the coefficient 1.

1.
$$\begin{aligned} x_1 + x_2 &= 3 \\ -2x_1 + 7x_2 &= 9 \end{aligned}$$

2.
$$\begin{aligned} 2x_1 + 3x_2 &= 6 \\ 4x_1 - x_2 &= 7 \end{aligned}$$

3.
$$\begin{aligned} x_1 + 2x_2 - x_3 &= 1 \\ x_1 + x_2 + 2x_3 &= 2 \\ -2x_1 + x_2 &= 4 \end{aligned}$$

4.
$$\begin{aligned} x_2 + x_3 &= 4 \\ x_1 - x_2 + 2x_3 &= 1 \\ 2x_1 + x_2 - x_3 &= 6 \end{aligned}$$

5.
$$\begin{aligned} x_1 + x_2 &= 9 \\ x_1 - x_2 &= 7 \\ 3x_1 + x_2 &= 6 \end{aligned}$$

6.
$$\begin{aligned} x_1 + x_2 + x_3 - x_4 &= 1 \\ -x_1 + x_2 - x_3 + x_4 &= 3 \\ -2x_1 + x_2 + x_3 - x_4 &- 2 \end{aligned}$$

In Exercises 7-10, use elementary operations to solve the system of equations.

7.
$$\begin{aligned} x_1 + 2x_2 &= -5 \\ 2x_1 - x_2 &= 5 \end{aligned}$$

8.
$$\begin{aligned} x_2 &= 6 \\ x_1 + x_2 &= 4 \end{aligned}$$

9.
$$\begin{aligned} x_1 \quad - x_3 &= 5 \\ -2x_1 + x_2 + 2x_3 &= -6 \\ 2x_2 + 2x_3 &= -4 \end{aligned}$$

10.
$$\begin{aligned} x_1 + 2x_2 + 4x_3 &= -9 \\ x_2 + 2x_3 &= -3 \\ x_1 + 2x_2 + 5x_3 &= -11 \end{aligned}$$

11. Solve the system of equations given in Exercise 7 by using Gauss-Jordan elimination to simplify the augmented matrix.

13

12. Solve the system of equations given in Exercise 8 by using Gauss-Jordan elimination to simplify the augmented matrix.

13. Solve the system of equations given in Exercise 9 by using Gauss-Jordan elimination to simplify the augmented matrix.

14. Solve the system of equations given in Exercise 10 by using Gauss-Jordan elimination to simplify the augmented matrix.

15. Solve the system of equations

$$
\begin{aligned}
2x_2 &- x_3 = 1 \\
3x_1 + 11x_2 &- 5x_3 = 1 \\
x_1 + 3x_2 &- x_3 = 1
\end{aligned}
$$

by performing the sequence $R_1 \leftrightarrow R_3$, $R_2 - 3R_1$, $(\frac{1}{2})R_2$, $R_1 - 3R_2$, $R_3 - 2R_2$, $R_1 - 2R_3$, $R_2 + R_3$ of elementary row operations on the augmented matrix for the system.

16. Solve the system of equations

$$
\begin{aligned}
2x_2 &- x_3 = 10 \\
-3x_1 - x_2 &+ 3x_3 = 7 \\
2x_1 + 2x_2 &+ 2x_3 = 2
\end{aligned}
$$

by performing the sequence $R_1 \leftrightarrow R_3$, $(\frac{1}{2})R_1$, $R_2 + 3R_1$, $(\frac{1}{2})R_2$, $R_1 - R_2$, $R_3 - 2R_2$, $(\frac{-1}{7})R_3$, $R_1 + 2R_3$, $R_2 - 3R_3$ of elementary row operations on the augmented matrix for the system.

17. In each of (a)-(c) the given elementary row operation transforms a matrix A to the matrix B. In each case, give the elementary row operation that transforms B to A.

(a) $R_3 - 2R_1$ (b) $R_1 \leftrightarrow R_2$ (c) $4R_2$

18. The sequence $R_1 \leftrightarrow R_2$, $R_3 - 3R_1$, $(\frac{1}{2})R_2$, $R_1 + R_2$, $R_3 + 2R_2$ of elementary row operations transforms a matrix A to the matrix B. Exhibit a sequence of elementary row operations that transforms B to A.

1.3 CONSISTENT SYSTEMS WITH INFINITELY MANY SOLUTIONS

The last section introduced Gauss-Jordan elimination as a method for solving linear systems. In that introduction we restricted ourselves to looking at consistent systems with unique solutions.

In this section we look at consistent systems that have infinitely many solutions. Our focus will be on how to describe the solutions to a system when there are infinitely many of them.

Geometric representations of (2×2) linear systems

The graph of a linear equation in two unknowns is a line in the plane. Thus, a *system* of two linear equations in two unknowns can be represented geometrically by two lines in the plane. The solution to the system is found at the intersection of the two lines.

So, consider the general (2×2) linear system:

$$a_{11}x_1 + a_{12}x_2 = b_1$$
$$a_{21}x_1 + a_{22}x_2 = b_2$$

Let l_1 denote the line corresponding to the first equation and let l_2 denote the line corresponding to the second equation. From plane geometry we know there are three possibilities:

(a)   The lines l_1 and l_2 intersect at a single point (thus, the system has a unique solution).

(b)   The lines l_1 and l_2 are parallel (thus, the system has no solution).

(c)   The lines l_1 and l_2 coincide (thus, the system has infinitely many solutions).

Systems illustrating the three possibilities are given below and graphed in Figure 1:

(a) $x_1 + x_2 = 3$ (b) $x_1 + x_2 = 2$ (c) $x_1 + x_2 = 2$
$\ x_1 - x_2 = 1$ $\ x_1 + x_2 = 1$ $\ 2x_1 + 2x_2 = 4$

Note, in the case of system (c), that both equations have the same graph. Thus every point on l_1 is also a point on l_2. Consequently, we see that system (c) has infinitely many solutions.

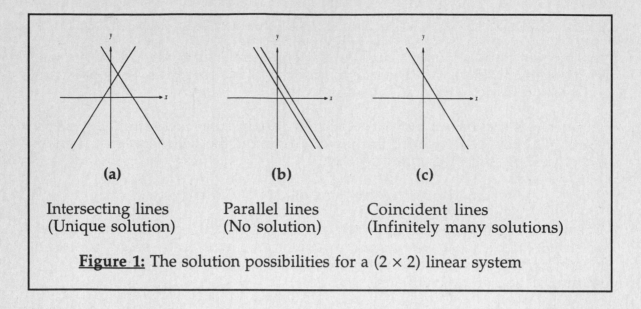

(a)

(b)

(c)

Intersecting lines
(Unique solution)

Parallel lines
(No solution)

Coincident lines
(Infinitely many solutions)

Figure 1: The solution possibilities for a (2×2) linear system

Describing solutions when there are infinitely many of them

When a linear system has just one solution, as in case **(a)** in Figure 1, there is no problem describing the solution—for case **(a)** the solution is $x_1 = 2$, $x_2 = 1$. Similarly, we have no problem with describing a solution if there is no solution to describe, as in case **(b)**.

So, let us turn our attention to the linear system **(c)**:

$$
\begin{aligned}
x_1 + x_2 &= 2 \\
2x_1 + 2x_2 &= 4
\end{aligned}
$$

We solve the system by reducing the augmented system as follows

$$
\begin{bmatrix} 1 & 1 & 2 \\ 2 & 2 & 4 \end{bmatrix} \xrightarrow{R_2 - 2R_1} \begin{bmatrix} 1 & 1 & 2 \\ 0 & 0 & 0 \end{bmatrix}
$$

Consequently, the system reduces to the single equation

$$
x_1 + x_2 = 2.
$$

Solving the above equation for x_1 yields an <u>algebraic description</u> of the set of solutions to system **(c)**, namely

(1) $$x_1 = 2 - x_2.$$

The equation above is called the **_general solution_** to system **(c)**. The variable x_2 is referred to as an **_independent variable_**, and the variable x_1 is the **_dependent variable_**.

Specific numerical solutions of system **(c)**, called *__particular solutions__*, are found by assigning values to the independent variable and then solving to find the corresponding value of the dependent variable x_1. For example, setting $x_2 = 1$ yields $x_1 = 1$. Other particular solutions, obtained by assigning specific values to x_2, include $x_1 = 3$, $x_2 = -1$; $x_1 = 2$, $x_2 = 0$; and $x_1 = 0$, $x_2 = 2$.

Finally, note from equation (1) that the solution to system **(c)** has one independent variable. We will say, informally, that system **(c)** has a *__one-dimensional__* solution set.

Two examples

Rectangular systems are a common source for systems having infinitely many solutions. Our next example illustrates this fact and also shows that Gauss-Jordan elimination can be used on a rectangular system as well as on square systems.

__Example 1__   Use Gauss-Jordan elimination to solve the following (2×3) system

$$x_1 + 3x_2 + 2x_3 = -2$$
$$2x_1 + 7x_2 + 5x_3 = -5$$

__Solution:__   The augmented matrix for the system can be reduced as follows:

$$\begin{bmatrix} 1 & 3 & 2 & -2 \\ 2 & 7 & 5 & -5 \end{bmatrix} \xrightarrow{R_2 - 2R_1} \begin{bmatrix} 1 & 3 & 2 & -2 \\ 0 & 1 & 1 & -1 \end{bmatrix} \xrightarrow{R_1 - 3R_2} \begin{bmatrix} 1 & 0 & -1 & 1 \\ 0 & 1 & 1 & -1 \end{bmatrix}$$

Reading from the final augmented matrix above, the original system is equivalent to the simplified system:

$$x_1 \quad\quad -x_3 = 1$$
$$x_2 + x_3 = -1$$

Solving for the leading variables in each equation, we obtain the general solution

$$x_1 = 1 + x_3$$
$$x_2 = -1 - x_3$$

∎

Note: We wish to make several comments about the system of two eqauations in three unknowns in Example 1. First of all, the graph of each of the equations in the system is a plane in three-space. Therefore, the solution of the system corresponds to the intersection of the two planes. From geometry, we know that two planes are either parallel, coincident, or intersect in a line. For the system in Example 1, the planes intersect in a line and the solution to Example 1 gives a parametric description of the line, namely:

$$x_1 = 1 + t$$
$$x_2 = -1 - t$$
$$x_3 = t, \qquad\qquad -\ddot{\text{I}} < t < \ddot{\text{I}}$$

Our final example can also be interpreted in terms of the intersection of planes. The example also illustrates a system with a two-dimensional solution set.

Example 2 Solve the (3×3) linear system

$$x_1 - 2x_2 + x_3 = 4$$
$$-2x_1 + 4x_2 - 2x_3 = -8$$
$$3x_1 - 6x_2 + 3x_3 = 12$$

Solution: We reduce the augmented matrix as follows:

$$\begin{bmatrix} 1 & -2 & 1 & 4 \\ -2 & 4 & -2 & -8 \\ 3 & -6 & 3 & 12 \end{bmatrix} \xrightarrow[\text{R_3-3R_1}]{\text{R_2+2R_1}} \begin{bmatrix} 1 & -2 & 1 & 4 \\ 0 & 0 & 0 & 0 \\ 0 & 0 & 0 & 0 \end{bmatrix}.$$

Thus, the system of three equations reduces to the single equation

(2) $$\qquad\qquad x_1 - 2x_2 + x_3 = 4.$$

The general solution is found by solving for the leading variable x_1:

$$x_1 = 4 + 2x_2 - x_3.$$

■

Note: The graph of each of the three equations in Example 2 is a plane. Thus, the solution of the linear system is the intersection of the three planes. Under the reduction process, the system was transformed to an equivalent single equation, equation (2). This means, in geometric terms, that the graphs of the three equations are coincident planes (that is, the same plane). Also, observe that the solution set found in Example 2 is two dimensional.

The general solution

To summarize, a system of equations may have infintely many solutions. Those solutions are described algebraically by a _general solution_ in which certain of the variables, designated as _dependent variables_, are expressed in terms of the remaining _independent variables_. Particular (numerical) solutions can obtained by assigning values to the independent variables. A measure of the size of the solution set is its _dimension_, which equals the number of independent variables in the general solution.

Exercises 1.3

In Exercises 1-4, determine whether the system has a unique solution, no solution, or infinitely many solutions by sketching a graph for each equation.

1.
$$\begin{aligned} 2x + y &= 5 \\ x - y &= 1 \end{aligned}$$

2.
$$\begin{aligned} 2x - y &= -1 \\ 2x - y &= 2 \end{aligned}$$

3.
$$\begin{aligned} 3x + 2y &= 6 \\ -6x - 4y &= -12 \end{aligned}$$

4.
$$\begin{aligned} 2x + y &= 5 \\ x - y &= 1 \\ x + 3y &= 9 \end{aligned}$$

In Exercises 5-7, determine whether the given (2×3) system of linear equations represents coincident planes (that is, the same plane), two parallel planes, or two planes whose intersection is a line. In the latter case, give parametric equations for the line; that is, give equations of the form $x_1 = at + b$, $x_2 = ct + d$, $x_3 = et + f$.

5.
$$\begin{aligned} 2x_1 + x_2 + x_3 &= 3 \\ -2x_1 + x_2 - x_3 &= 1 \end{aligned}$$

6.
$$\begin{aligned} x_1 + 2x_2 - x_3 &= 2 \\ x_1 + x_2 + x_3 &= 3 \end{aligned}$$

7.
$$\begin{aligned} x_1 + 3x_2 - 2x_3 &= -1 \\ 2x_1 + 6x_2 - 4x_3 &= -2 \end{aligned}$$

In Exercises 8-10 you are given the augmented matrix for a reduced system of equations. In each case, exhibit the corresponding reduced system of equations, give the general solution for the system, and state which variables are independent.

8.
$$\begin{bmatrix} 1 & 0 & -3 & 1 \\ 0 & 1 & 2 & -5 \\ 0 & 0 & 0 & 0 \end{bmatrix}$$

9.
$$\begin{bmatrix} 1 & -2 & 0 & 4 \\ 0 & 0 & 1 & 0 \\ 0 & 0 & 0 & 0 \end{bmatrix}$$

10.
$$\begin{bmatrix} 1 & -4 & 3 & 5 \\ 0 & 0 & 0 & 0 \\ 0 & 0 & 0 & 0 \end{bmatrix}$$

11. In each of parts (a)-(c) below, exhibit a particular solution for the system found in Exercise 8 that satisfies the given constraint.

$$(\text{a}) \ x_3 = 1 \quad (\text{b}) \ x_3 = -1 \quad (\text{c}) \ x_3 = 0$$

12. In each of parts (a)-(c) below, exhibit a particular solution for the system found in Exercise 9 that satisfies the given constraint.

$$(\text{a}) \ x_2 = 1 \quad (\text{b}) \ x_2 = 2 \quad (\text{c}) \ x_2 = 0$$

13. In each of parts (a)-(c) below, exhibit a particular solution for the system found in Exercise 10 that satisfies the given constraint.

$$(\text{a}) \ x_2 = 1, \ x_3 = 0 \quad (\text{b}) \ x_2 = 0, \ x_3 = 1 \quad (\text{c}) \ x_2 - 1, \ x_3 - 1$$

14. Find the general solution for the system of equations

$$
\begin{array}{rcrcrcr}
x_1 & + & 2x_2 & + & 3x_3 & = & 0 \\
-x_1 & + & x_2 & + & 3x_3 & = & -6 \\
3x_1 & + & 4x_2 & + & 5x_3 & = & 4
\end{array}
$$

by performing the sequence R_2+R_1, R_3-3R_1, $(\frac{1}{3})R_2$, R_1-2R_2, R_3+2R_2 of elementary row operations on the augmented matrix for the system. State which variables are independent.

15. In each of parts (a)-(c) below, exhibit a particular solution for the system found in Exercise 14 that satisfies the given constraint.

$$(\text{a}) \ x_3 = -3 \quad (\text{b}) \ x_3 = 5 \quad (\text{c}) \ x_3 = 1$$

In Exercises 16-18, find the general solution of the given system of equations by using Gauss-Jordan elimination to simplify the augmented matrix. Exhibit three particular solutions for the given system.

16.
$$\begin{aligned} x_1 + 3x_2 &= -1 \\ -2x_1 - 6x_2 &= 2 \end{aligned}$$

17.
$$\begin{aligned} x_1 \qquad\; + 2x_3 &= -1 \\ 2x_1 + x_2 + x_3 &= 2 \\ x_1 + 2x_2 - 4x_3 &= 7 \end{aligned}$$

18.
$$\begin{aligned} x_1 - 3x_2 + 2x_3 &= -4 \\ -2x_1 + 6x_2 - 4x_3 &= 8 \\ 3x_1 - 9x_2 + 6x_3 &= -12 \end{aligned}$$

1.4 INCONSISTENT SYSTEMS

In Section 2 we considered linear systems having just one solution. Then, in Section 3, we considered systems that had infinitely many solutions. In this section we examine another possibility—the case when a linear system has no solutions at all.

Recognizing an inconsistent system

Recall that a linear system with no solution is called *inconsistent*. It is easy to construct inconsistent systems and inconsistent equations. For example, the equation

$$0x = 1$$

is obviously inconsistent since we can find no value for x such that zero times x is equal to 1.

Similarly, the following equation in two unknowns is inconsistent

$$0x_1 + 0x_2 = 1.$$

In general, the equation below has no solution

(1) $$0x_1 + 0x_2 + \cdots + 0x_n = 1.$$

As we will see, any linear system which has no solution can be reduced by Gauss-Jordan elimination to an equivalent system which has at least one equation of the form (1).

When we begin solving a linear system we usually have no way of knowing whether it will have a unique solution, infinitely many solutions, or no solutions at all. If the Gauss-Jordan process leads to an equation of the form (1), then the system is inconsistent. If Gauss-Jordan does not lead to such an equation, then the system is consistent.

The augmented matrix for an inconsistent system

Our first step in solving a linear system is forming the augmented matrix. If the system is inconsistent, then the Gauss-Jordan process will lead to a matrix having at least one row that corresponds to an equation of the form (1), a row of the form

$$[\,0, 0, \ldots, 0, 1\,].$$

The example below illustrates this point.

Example 1   Use Gauss-Jordan elimination to solve each of the linear systems

(a) $x_1 + x_2 = 1$
     $2x_1 + 2x_2 = 4$

(b) $x_1 + 2x_2 - x_3 = 1$
     $3x_1 + 6x_2 - 3x_3 = 4$

Solution: In terms of augmented matrices, the elimination proceeds as follows

System (a)

$$\begin{bmatrix} 1 & 1 & 1 \\ 2 & 2 & 4 \end{bmatrix} \xrightarrow{R_2 - 2R_1} \begin{bmatrix} 1 & 1 & 1 \\ 0 & 0 & 2 \end{bmatrix} \xrightarrow{(1/2)R_2} \begin{bmatrix} 1 & 1 & 1 \\ 0 & 0 & 1 \end{bmatrix} \xrightarrow{R_1 - R_2} \begin{bmatrix} 1 & 1 & 0 \\ 0 & 0 & 1 \end{bmatrix}.$$

Looking at the second row, we know that sytem **(a)** has no solution. That is, reading from the last augmented matrix above, we see that system **(a)** is equivalent to a system that has an impossible equation:

$$x_1 + x_2 = 0$$
$$0x_1 + 0x_2 = 1$$

System (b)

$$\begin{bmatrix} 1 & 2 & -1 & 1 \\ 3 & 6 & -3 & 4 \end{bmatrix} \xrightarrow{R_2 - 3R_1} \begin{bmatrix} 1 & 2 & -1 & 1 \\ 0 & 0 & 0 & 1 \end{bmatrix} \xrightarrow{R_1 - R_2} \begin{bmatrix} 1 & 2 & -1 & 0 \\ 0 & 0 & 0 & 1 \end{bmatrix}$$

From the second row of the reduced matrix, we see that system **(b)** is inconsistent. ∎

Note: For either of the systems in Example 1, we could have stopped the Gauss-Jordan elimination process sooner than we did. In each case, a single elementary row operation brought us to a row where the leading nonzero entry was in the last column. Any time we encounter such an augmented matrix, we know the corresponding system is inconsistent.

Example 2   Solve the system

$$x_1 + 2x_2 \qquad\quad = -5$$
$$2x_1 - x_2 + 5x_3 = -5$$
$$3x_1 + 4x_2 + 2x_3 = -3$$

Solution: Gauss-Jordan elimination proceeds as follows

$$\begin{bmatrix} 1 & 2 & 0 & -5 \\ 2 & -1 & 5 & -5 \\ 3 & 4 & 2 & -3 \end{bmatrix} \xrightarrow[R_3-3R_1]{R_2-2R_1} \begin{bmatrix} 1 & 2 & 0 & -5 \\ 0 & -5 & 5 & 5 \\ 0 & -2 & 2 & 12 \end{bmatrix} \xrightarrow{-(1/5)R_2} \begin{bmatrix} 1 & 2 & 0 & -5 \\ 0 & 1 & -1 & -1 \\ 0 & -2 & 2 & 12 \end{bmatrix}$$

$$\xrightarrow[R_3+2R_2]{R_1-2R_2} \begin{bmatrix} 1 & 0 & 2 & -3 \\ 0 & 1 & -1 & -1 \\ 0 & 0 & 0 & 10 \end{bmatrix}$$

We could continue the Gauss-Jordan process, but the third row of the last matrix tells us that the system is inconsistent. ∎

Example 3 Each of the following is an augmented matrix for a system of linear equations. Solve the associated system or state that it is inconsistent. If the system has infinitely many solutions, give the dimension of the solution set. (Each matrix has been reduced as far as it can be, so there is no need to use Gauss-Jordan elimination.)

(a) $\begin{bmatrix} 1 & 0 & 2 \\ 0 & 0 & 1 \end{bmatrix}$
(b) $\begin{bmatrix} 1 & 1 & 0 & 2 \\ 0 & 0 & 1 & 3 \\ 0 & 0 & 0 & 0 \end{bmatrix}$
(c) $\begin{bmatrix} 1 & 0 & 2 \\ 0 & 1 & 3 \\ 0 & 0 & 0 \end{bmatrix}$

(d) $\begin{bmatrix} 1 & 1 & 0 & 0 & 2 \\ 0 & 0 & 1 & 0 & 1 \end{bmatrix}$
(e) $\begin{bmatrix} 1 & 0 & 0 & 0 \\ 0 & 1 & 0 & 0 \\ 0 & 0 & 1 & 0 \\ 0 & 0 & 0 & 1 \end{bmatrix}$

Solution:

(a) The system is inconsistent since the second equation is $0x_1 + 0x_2 = 1$.

(b) The system is consistent. The solution is $x_1 = 2 - x_2$, $x_3 = 3$, x_2 arbitrary. The solution set is one-dimensional.

(c) The system is consistent. The unique solution is $x_1 = 2$, $x_2 = 3$.

(d) The system is consistent. The solution is $x_1 = 2 - x_2$, $x_3 = 1$, x_2 and x_4 arbitrary. The solution set is two-dimensional.

(e) The system is inconsistent since the fourth equation is $0x_1 + 0x_2 + 0x_3 = 1$. ∎

A geometric interpretation for (3 × 3) systems

In Section 3 we saw a figure illustrating the three solution possibilities for a (2 × 2) linear system. A similar interpretation can be made for a (3 × 3) system. In particular, consider the general (3 × 3) linear system

$$a_{11}x_1 + a_{12}x_2 + a_{13}x_3 = b_1$$
$$a_{21}x_1 + a_{22}x_2 + a_{23}x_3 = b_2$$
$$a_{31}x_1 + a_{32}x_2 + a_{33}x_3 = b_3$$

The graph of each equation is a plane in three-dimensional space. As can be seen from Figure 1, there are three possibilities for the (3 × 3) system:

(a)   The system has a unique solution.
(b)   The system has no solution.
(c) - (d)   The system has infinitely many solutions.

In Figure 1(c) the solution set is one-dimensional. Figure 1(d) illustrates the case where all three planes are coincident, giving a two-dimensional solution set.

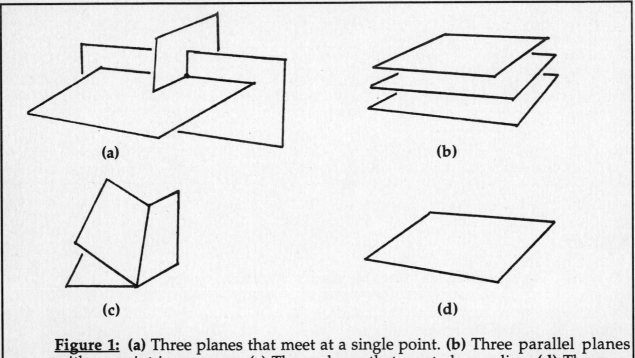

(a)   (b)

(c)   (d)

Figure 1: **(a)** Three planes that meet at a single point. **(b)** Three parallel planes with no point in common. **(c)** Three planes that meet along a line. **(d)** Three coincident planes.

Exercises 1.4

In Exercises 1-4 you are given the augmented matrix for a reduced system of equations. Either provide a solution for the corresponding system of equations or state that the system is inconsistent.

1. $\begin{bmatrix} 1 & 2 & 0 \\ 0 & 0 & 1 \end{bmatrix}$
2. $\begin{bmatrix} 1 & -3 & 0 & 4 \\ 0 & 0 & 1 & 0 \end{bmatrix}$

3. $\begin{bmatrix} 1 & 0 & 0 \\ 0 & 1 & 0 \\ 0 & 0 & 1 \end{bmatrix}$
4. $\begin{bmatrix} 1 & 0 & 0 & 0 \\ 0 & 1 & 2 & 3 \end{bmatrix}$

In Exercises 5-8, the given system of equations is inconsistent. Verify this fact by using Gauss-Jordan elimination to simplify the augmented matrix.

5.
$$\begin{aligned} x_1 + x_2 &= 3 \\ -2x_1 - 2x_2 &= 4 \end{aligned}$$

6.
$$\begin{aligned} 2x_1 - 3x_2 &= 4 \\ 6x_1 - 9x_2 &= 4 \end{aligned}$$

7.
$$\begin{aligned} x_1 + 2x_2 + x_3 &= 3 \\ x_1 - x_2 + x_3 &= 1 \\ -2x_1 - 4x_2 - 2x_3 &= 4 \end{aligned}$$

8.
$$\begin{aligned} x_1 - 2x_2 - x_3 &= 1 \\ x_1 - x_2 + x_3 &= 0 \\ 2x_1 \quad\quad + 6x_3 &= 1 \end{aligned}$$

In Exercises 9-12, find all values of a for which the system has no solution.

9.
$$\begin{aligned} x_1 + x_2 &= 5 \\ 2x_1 + ax_2 &= 4 \end{aligned}$$

10.
$$\begin{aligned} x_1 + 2x_2 &= -3 \\ ax_1 - 2x_2 &= 5 \end{aligned}$$

11.
$$\begin{aligned} x_1 + 3x_2 &= 4 \\ 2x_1 + 6x_2 &= a \end{aligned}$$

12.
$$\begin{aligned} 2x_1 + 4x_2 &= a \\ 3x_1 + 6x_2 &= 5 \end{aligned}$$

In Exercises 13-16, solve the system of equations or verify that it is inconsistent.

$$\begin{aligned} x_1 \quad\quad + \ 3x_3 &= \ 2 \\ 13. \quad 2x_1 + \ x_2 + \ 2x_3 &= \ 5 \\ -x_1 - \ 2x_2 + \ 5x_3 &= -1 \end{aligned}$$

$$\begin{aligned} x_1 + \ 5x_2 \quad\quad &= \ -4 \\ 14. \quad -2x_1 - \ 10x_2 + \ x_3 &= \ 8 \\ 3x_1 + \ 15x_2 + \ 2x_3 &= -12 \end{aligned}$$

$$\begin{aligned} x_1 \quad\quad + \ 2x_3 &= -3 \\ 15. \quad\quad x_2 - \ x_3 &= \ 2 \\ x_1 + \ 3x_2 + \ x_3 &= \ 3 \end{aligned}$$

$$\begin{aligned} x_1 - \ 2x_2 + \ 3x_3 &= \ 5 \\ 16. \quad 2x_1 - \ 4x_2 + \ 6x_3 &= \ 13 \\ 3x_1 - \ 6x_2 + \ 9x_3 &= \ 19 \end{aligned}$$

1.5 REDUCED ECHELON FORM

The general procedure for solving a system of linear equations can be stated as follows:

Step 1: Form the augmented matrix for the system.

Step 2: Use elementary row operations to reduce the augmented matrix to a simpler form.

Step 3: Solve the simpler but equivalent system represented by the reduced matrix.

The entire process is diagrammed in Figure 1.

$$\begin{bmatrix} \text{Given system} \\ \text{of equations} \end{bmatrix} \rightarrow \begin{bmatrix} \text{Augmented} \\ \text{matrix} \end{bmatrix} \rightarrow \begin{bmatrix} \text{Reduced} \\ \text{matrix} \end{bmatrix} \rightarrow \begin{bmatrix} \text{Reduced system} \\ \text{of equations} \end{bmatrix} \rightarrow \begin{bmatrix} \text{Solution of} \\ \text{equivalent system} \end{bmatrix}$$

Figure 1: Procedure for solving a system of linear equations

The objective of the reduction step (represented by the middle block in Figure 1) is to simplify the augmented matrix to the point where the solution is easy to find. An obvious question then arises: "What do we mean when we say 'simplify the augmented matrix'? What sort of simplifications are we aiming for?" An example may help to guide us.

Example 1 Solve the following system of equations

$$\begin{aligned} x_1 + x_2 \quad\quad\quad + 3x_4 &= -2 \\ 2x_1 + 2x_2 + x_3 + 10x_4 &= -1 \\ -x_1 - x_2 + 2x_3 + 5x_4 &= 8 \end{aligned}$$

Solution: Following the steps outlined in Figure 1, we first create the augmented matrix and then begin simplifying it:

$$\begin{bmatrix} 1 & 1 & 0 & 3 & -2 \\ 2 & 2 & 1 & 10 & -1 \\ -1 & -1 & 2 & 5 & 8 \end{bmatrix} \xrightarrow[\substack{R_2-2R_1 \\ R_3+R_1}]{} \begin{bmatrix} 1 & 1 & 0 & 3 & -2 \\ 0 & 0 & 1 & 4 & 3 \\ 0 & 0 & 2 & 8 & 6 \end{bmatrix} \xrightarrow[R_3-2R_2]{} \begin{bmatrix} 1 & 1 & 0 & 3 & -2 \\ 0 & 0 & 1 & 4 & 3 \\ 0 & 0 & 0 & 0 & 0 \end{bmatrix}$$

Now, what do we do? We can perform more row operations on the third matrix, but it soon becomes clear that no additional simplification is occurring.

In fact, as we will see in the next subsection, the third matrix is already in the simplest possible form, a form called "reduced echelon form." Note that the third matrix represents the following system of equations:

$$x_1 + x_2 \quad + 3x_4 = -2$$
$$x_3 + 4x_4 = 3$$

This system meets our criterion for simplicity since the solution can be found immediately:

$$x_1 = -2 - x_2 - 3x_4$$
$$x_3 = 3 \qquad - 4x_4$$

■

Reduced echelon form

We are now ready to describe reduced echelon form. This is the target we are aiming for when want to simplify an augmented matrix.

DEFINITION 1   A matrix is in *__reduced echelon form__* provided:

1.   All rows that consist entirely of zeros are grouped together at the bottom of the matrix.

2.   In any nonzero row, the first nonzero entry (counting from the left) is a 1.

3.   If a row after the first row contains nonzero entries, then the first nonzero entry in that row appears in a column to the right of the first nonzero entry in the preceding row.

4.   The first nonzero entry in any row is the only nonzero entry in its column.

At first glance, the definition of reduced echelon form might seem to be overly complicated, especially items 3 and 4. The concept is really quite simple, however.

If a matrix is in reduced echelon form, then the nonzero entries form a staircase-like pattern. For instance, Figure 2 shows all possible patterns for a nonzero (3 × 3) matrix in reduced echelon form.

$$\begin{bmatrix} 1 & 0 & 0 \\ 0 & 1 & 0 \\ 0 & 0 & 1 \end{bmatrix} \begin{bmatrix} 1 & 0 & \times \\ 0 & 1 & \times \\ 0 & 0 & 0 \end{bmatrix} \begin{bmatrix} 1 & \times & 0 \\ 0 & 0 & 1 \\ 0 & 0 & 0 \end{bmatrix} \begin{bmatrix} 1 & \times & \times \\ 0 & 0 & 0 \\ 0 & 0 & 0 \end{bmatrix} \begin{bmatrix} 0 & 1 & 0 \\ 0 & 0 & 1 \\ 0 & 0 & 0 \end{bmatrix} \begin{bmatrix} 0 & 1 & \times \\ 0 & 0 & 0 \\ 0 & 0 & 0 \end{bmatrix} \begin{bmatrix} 0 & 0 & 1 \\ 0 & 0 & 0 \\ 0 & 0 & 0 \end{bmatrix}$$

Figure 2: The seven possible forms for a nonzero (3×3) matrix in reduced echelon form. The entries marked $\times$ may be either zero or nonzero.

Two numerical examples of matrices in reduced echelon form are displayed below. Note that the leading 1 in each row has nothing but 0's above and below it (recall item 4 from the definition of reduced echelon form). Also note the characteristic staircase-like pattern of the nonzero entries.

(1)
$$A = \begin{bmatrix} 1 & 0 & 0 & 2 \\ 0 & 1 & 0 & -1 \\ 0 & 0 & 1 & 3 \end{bmatrix} \, , \quad B = \begin{bmatrix} 1 & 2 & 0 & 1 & -1 \\ 0 & 0 & 1 & 3 & 4 \\ 0 & 0 & 0 & 0 & 0 \end{bmatrix}$$

Matrix B illustrates items 1 - 4 of Definition 1. First, all the zero rows are at the bottom of the matrix. Second, the leading entry in each nonzero row is a 1. Third, the leading 1 in row two is to the right of the leading 1 in row one. Fourth, in rows one and two, there are nothing but 0s above and below the leading 1.

Our next example illustrates how easy it is to solve a system of equations when its augmented matrix has been transformed to reduced echelon form.

Example 2: Let A and B in (1) each be an augmented matrix for a system. Solve each of the systems.

Solution: The system represented by A is:

$$
\begin{aligned}
x_1 \quad\quad\quad &= 2 \\
x_2 \quad\quad &= -1 \\
x_3 &= 3
\end{aligned}
$$

The solution to this system is $x_1 = 2$, $x_2 = -1$, $x_3 = 3$.

The system represented by B is:

$$
\begin{aligned}
x_1 + 2x_2 \quad\quad x_4 &= -1 \\
x_3 + 3x_4 &= 4
\end{aligned}
$$

The solution to this system is $x_1 = -1 - 2x_2 - x_4$, $x_3 = 4 - 3x_4$. ∎

Example 3: Determine which of the following matrices are in reduced echelon form.

$$A = \begin{bmatrix} 1 & 0 & 0 \\ 2 & 1 & 0 \\ 3 & -4 & 1 \end{bmatrix} \quad B = \begin{bmatrix} 1 & 3 & 2 \\ 0 & 1 & 1 \\ 0 & 0 & 1 \end{bmatrix} \quad C = \begin{bmatrix} 0 & 1 & -1 & 0 \\ 0 & 0 & 0 & 1 \\ 0 & 0 & 0 & 0 \end{bmatrix} \quad D = \begin{bmatrix} 1 & 2 & 0 & 0 & 4 \\ 0 & 0 & 1 & 0 & 3 \\ 0 & 0 & 0 & 1 & 0 \end{bmatrix}$$

$$E = \begin{bmatrix} 1 \\ 0 \\ 0 \end{bmatrix} \qquad F = \begin{bmatrix} 0 \\ 0 \\ 1 \end{bmatrix} \qquad G = \begin{bmatrix} 1 & 0 & 0 \end{bmatrix} \qquad H = \begin{bmatrix} 0 & 0 & 1 \end{bmatrix}$$

Solution: The matrices C, D, E, G, and H are in reduced echelon form. ∎

Note that the matrix B in Example 3 satisfies the first three conditions in the definition for reduced echelon form, but not the fourth. Such a matrix is said to be in *echelon form*.

Exercises 1.5

1. The matrix

$$\begin{bmatrix} 1 & 0 & \times \\ 0 & 1 & \times \end{bmatrix}$$

is one configuration for a nonzero (2×3) matrix in reduced echelon form, where the entries marked $\times$ can be zero or nonzero. Display all five remaining configurations that are possible.

2. Display all possible configurations for a nonzero (3×2) matrix in reduced echelon form.

In Exercises 3-10, either state that the given matrix is in reduced echelon form or use elementary row operations to transform it into reduced echelon form.

3. $\begin{bmatrix} 1 & 2 \\ 0 & 1 \end{bmatrix}$ 4. $\begin{bmatrix} 1 & 2 & -1 \\ 0 & 1 & 3 \end{bmatrix}$

5. $\begin{bmatrix} 0 & 1 & -3 \\ 0 & 0 & 0 \end{bmatrix}$ 6. $\begin{bmatrix} 1 & 4 & 0 & 0 \\ 0 & 0 & 1 & 0 \\ 0 & 0 & 0 & 1 \end{bmatrix}$

7. $\begin{bmatrix} 1 & 3 & 2 & 1 \\ 0 & 1 & 4 & 2 \\ 0 & 0 & 1 & 1 \end{bmatrix}$ 8. $\begin{bmatrix} 0 & 0 & 0 & 1 \\ 0 & 1 & 0 & 0 \\ 1 & 0 & 0 & 0 \end{bmatrix}$

9. $\begin{bmatrix} 1 & -1 & 5 & 0 \\ 0 & 0 & 0 & 1 \\ 0 & 0 & 0 & 0 \end{bmatrix}$ 10. $\begin{bmatrix} 1 & 2 & -1 & -2 \\ 0 & 2 & -2 & -3 \\ 0 & 0 & 0 & 1 \end{bmatrix}$

In Exercises 11-16, solve the given system of equations by transforming the augmented matrix to reduced echelon form.

11. $\begin{aligned} x_1 + 2x_2 &= -5 \\ 3x_1 + 8x_2 &= -23 \end{aligned}$

33

12.
$$\begin{aligned} x_1 \qquad\quad -5x_3 &= 3 \\ -x_1 + 2x_2 + 9x_3 &= -5 \\ 2x_1 + 3x_2 - 4x_3 &= 3 \end{aligned}$$

13.
$$\begin{aligned} -x_2 - 3x_3 &= 2 \\ x_1 \qquad\quad - 4x_3 &= 3 \\ 2x_1 + 3x_2 + x_3 &= 5 \end{aligned}$$

14.
$$\begin{aligned} x_1 - 3x_2 + 2x_3 + x_4 &= 14 \\ -x_1 + 3x_2 + x_3 - 7x_4 &= 4 \\ 2x_1 - 6x_2 + 2x_3 + 6x_4 &= 16 \end{aligned}$$

15.
$$\begin{aligned} x_1 \qquad\quad + 2x_3 + x_4 &= -1 \\ 2x_1 + 3x_2 + 4x_3 + 2x_4 &= -2 \\ -x_1 + 6x_2 - 2x_3 + x_4 &= -9 \end{aligned}$$

16.
$$\begin{aligned} x_1 \qquad\quad + 2x_3 &= 3 \\ -x_1 + x_2 - 2x_3 &= -5 \\ -2x_2 + 3x_3 &= 4 \\ x_1 - x_2 - x_3 &= 5 \end{aligned}$$

1.6 TRANSFORMING A MATRIX TO REDUCED ECHELON FORM

In this section we show how to use Gauss-Jordan elimination to transform a matrix to reduced echelon form. We have discussed Gauss-Jordan elimination in earlier sections, but now we want to list the steps in an unambiguous way.

Every matrix can be transformed to reduced echelon form

The theorem that follows says every matrix A can be transformed to reduced echelon form. In terms of solving a system of linear equations, this theorem says the augmented matrix can always be simplified to the point where the solution is easy to obtain.

THEOREM 1   Let A be an $(m \times n)$ matrix. There is a unique $(m \times n)$ matrix B such that:

(a) A can be transformed to B by a sequence of elementary row operations.

(b) B is in reduced echelon form.

The following steps show how to transform any matrix A to reduced echelon form. These steps constitute an informal proof of parts **(a)** and **(b)** of Theorem 1.

Transforming a matrix to reduced echelon form

Step 1:   Locate the first (leftmost) column that contains a nonzero entry.

Step 2:   If necessary, interchange the first row with another row so that the first nonzero column has a nonzero entry in the first row.

Step 3:   If a denotes the leading nonzero entry in row one, multiply each entry in row one by $1/a$. (Thus, the leading nonzero entry in row one is a 1.)

Step 4:   Add appropriate multiples of row one to each of the remaining rows so that every entry below the leading 1 in row one is a 0.

Step 5:   Temporarily ignore the first row of this matrix and repeat Steps 1-4 on the submatrix that remains. Stop the process when the resulting matrix is in echelon form.

Step 6: Having reached echelon form in Step 5, continue on to reduced echelon form as follows: Add multiples of each nonzero row to the rows above in order to zero all entries above the leading 1.

The following example illustrates an application of the six-step process listed above. When doing a small problem by hand, however, it is customary to alter the steps slightly—instead of going all the way to echelon form (sweeping from left to right) and then going from echelon to reduced echelon form (sweeping from bottom to top), it is customary to make a single pass (moving from left to right) introducing 0's above as well as below the leading 1. Example 1 demonstrates this single-pass variation.

Example 1: Use elementary row operations to transform the following matrix to reduced echelon form

$$A = \begin{bmatrix} 0 & 0 & 0 & 0 & 2 & 8 & 4 \\ 0 & 0 & 0 & 1 & 3 & 11 & 9 \\ 0 & 3 & -12 & -3 & -9 & -24 & -33 \\ 0 & -2 & 8 & 1 & 6 & 17 & 21 \end{bmatrix}.$$

Solution: The following row operations will transform A to reduced echelon form.

$R_1 <\!-\!\!> R_3, (1/3)R_1$:   (Introduce a leading 1 into the first row of the first nonzero column)

$$\begin{bmatrix} 0 & 1 & -4 & -1 & -3 & -8 & -11 \\ 0 & 0 & 0 & 1 & 3 & 11 & 9 \\ 0 & 0 & 0 & 0 & 2 & 8 & 4 \\ 0 & -2 & 8 & 1 & 6 & 17 & 21 \end{bmatrix}$$

$R_4 + 2R_1$:   (Introduce 0's below the leading 1 in row 1)

$$\begin{bmatrix} 0 & 1 & -4 & -1 & -3 & -8 & -11 \\ 0 & 0 & 0 & 1 & 3 & 11 & 9 \\ 0 & 0 & 0 & 0 & 2 & 8 & 4 \\ 0 & 0 & 0 & -1 & 0 & 1 & -1 \end{bmatrix}$$

$R_1 + R_2, R_4 + R_2$:   (Introduce 0's above and below the leading 1 in row 2)

$$\begin{bmatrix} 0 & 1 & -4 & 0 & 0 & 3 & -2 \\ 0 & 0 & 0 & 1 & 3 & 11 & 9 \\ 0 & 0 & 0 & 0 & 2 & 8 & 4 \\ 0 & 0 & 0 & 0 & 3 & 12 & 8 \end{bmatrix}$$

(1/2)R_3:　　　　　　　　　　(Introduce a leading 1 into row 3)

$$\begin{bmatrix} 0 & 1 & -4 & 0 & 0 & 3 & -2 \\ 0 & 0 & 0 & 1 & 3 & 11 & 9 \\ 0 & 0 & 0 & 0 & 1 & 4 & 2 \\ 0 & 0 & 0 & 0 & 3 & 12 & 8 \end{bmatrix}$$

$R_2 - 3R_3$, $R_4 - 3R_3$:　　　　(Introduce 0's above and below the leading 1 in row 3)

$$\begin{bmatrix} 0 & 1 & -4 & 0 & 0 & 3 & -2 \\ 0 & 0 & 0 & 1 & 0 & -1 & 3 \\ 0 & 0 & 0 & 0 & 1 & 4 & 2 \\ 0 & 0 & 0 & 0 & 0 & 0 & 2 \end{bmatrix}$$

(1/2)R_4:　　　　　　　　　　(Introduce a leading 1 into row 4)

$$\begin{bmatrix} 0 & 1 & -4 & 0 & 0 & 3 & -2 \\ 0 & 0 & 0 & 1 & 0 & -1 & 3 \\ 0 & 0 & 0 & 0 & 1 & 4 & 2 \\ 0 & 0 & 0 & 0 & 0 & 0 & 1 \end{bmatrix}$$

$R_1 + 2R_4$, $R_2 - 3R_4$, $R_3 - 2R_4$:　　(Introduce 0's above the leading 1 in row 4)

$$\begin{bmatrix} 0 & 1 & -4 & 0 & 0 & 3 & 0 \\ 0 & 0 & 0 & 1 & 0 & -1 & 0 \\ 0 & 0 & 0 & 0 & 1 & 4 & 0 \\ 0 & 0 & 0 & 0 & 0 & 0 & 1 \end{bmatrix}$$

■

Exercises 1.6

In Exercises 1-6, use elementary row operations to transform the given matrix to reduced echelon form.

1. $\begin{bmatrix} 3 & 11 \\ 2 & 8 \end{bmatrix}$

2. $\begin{bmatrix} 2 & -6 & -1 \\ 1 & -3 & -2 \end{bmatrix}$

3. $\begin{bmatrix} 5 & -4 & 23 \\ 2 & -1 & 8 \\ -3 & 4 & -17 \end{bmatrix}$

4. $\begin{bmatrix} 4 & -12 & -7 & 21 \\ -2 & 6 & 7 & -15 \\ 3 & -9 & -5 & 15 \end{bmatrix}$

5. $\begin{bmatrix} 3 & -4 & -15 \\ -4 & -4 & -8 \\ 1 & 2 & 5 \\ 1 & -1 & -4 \end{bmatrix}$

6. $\begin{bmatrix} 2 & 12 & 5 & 13 & -16 \\ 3 & 18 & 8 & 21 & -26 \\ 1 & 6 & 3 & 8 & -10 \\ 2 & 12 & 7 & 19 & -24 \end{bmatrix}$

In Exercises 7-12, solve the given system of equations by transforming the augmented matrix to reduced echelon form.

7.
$$\begin{aligned}
2x_1 + 3x_2 - 4x_3 &= 3 \\
x_1 - 2x_2 - 2x_3 &= -2 \\
-x_1 + 16x_2 + 2x_3 &= 16
\end{aligned}$$

8.
$$\begin{aligned}
x_1 + x_2 - x_3 &= 1 \\
2x_1 - x_2 + 7x_3 &= 8 \\
-x_1 + x_2 - 5x_3 &= -5
\end{aligned}$$

9.
$$\begin{aligned}
x_1 - x_2 - x_3 &= 1 \\
x_1 \quad\quad + x_3 &= 2 \\
x_2 + 2x_3 &= 3
\end{aligned}$$

10.
$$\begin{aligned}
x_1 + x_2 \quad\quad\quad\quad - x_5 &= 1 \\
x_2 + 2x_3 + x_4 + 3x_5 &= 1 \\
x_1 \quad\quad - x_3 + x_4 + x_5 &= 0
\end{aligned}$$

11.
$$\begin{aligned}
x_1 + x_2 &= 1 \\
x_1 - x_2 &= 3 \\
2x_1 + x_2 &= 3
\end{aligned}$$

38

$$12. \quad \begin{aligned} x_1 &+ 2x_2 &= 1 \\ 2x_1 &+ 4x_2 &= 2 \\ -x_1 &- 2x_2 &= -1 \end{aligned}$$

13. Any circle in the xy-plane has an equation of the form $x^2 + ax + y^2 + by + c = 0$. Find the center and the radius of the circle passing through the points $P(1,0)$, $Q(2,3)$, and $R(1,1)$.

14. Find the three numbers whose sum is 34, where the sum of the first and second is 7 and the sum of the second and third is 22.

15. A zoo charges \$6 for adults, \$3 for students, and \$0.50 for children. One morning, 79 people enter and pay a total of \$207. Determine the possible numbers of adults, students, and children.

1.7 HOW MANY SOLUTIONS CAN A LINEAR SYSTEM HAVE?

Because we have the concept of reduced echelon form, we are ready to answer the question posed in this section's title. In particular, we will prove that the _only_ possibilities for a system of linear equations are:

(a) The system might have no solution.
(b) The system might have exactly one solution.
(c) The system might have infinitely many solutions.

We have talked about this question before. In earlier sections, however, we only treated special cases: systems with two unknowns (see Figure 1, Section 3) and systems with three unknowns (see Figure 1, Section 4). Now that we have the concept of reduced echelon form, we can examine the general case of m equations in n unknowns.

Solution possibilities for nonlinear systems

Before taking up solution possibilities for linear systems, we briefly contrast linear systems with nonlinear systems.

The basic difference between solution possibilities for linear and nonlinear systems is that a nonlinear system might have exactly 2 solutions, or exactly 3 solutions, or (say) exactly 17 solutions. By contrast, if a linear system has exactly k solutions, then it must be that $k = 1$.

Example 1 Graph each equation in the following system of nonlinear equations. From your graph, determine how many solutions the system has:

$$x^2 + y^2 = 9$$
$$xy = 1$$

Solution: The solution set for the first equation is a circle of radius 3, centered at the origin (see Figure 1). The solution set for the second equation is a hyperbola (see Figure 1). The solution of the system is the intersection of the two solution sets. As can be seen from Figure 1, this particular nonlinear system has exactly four solutions. ∎

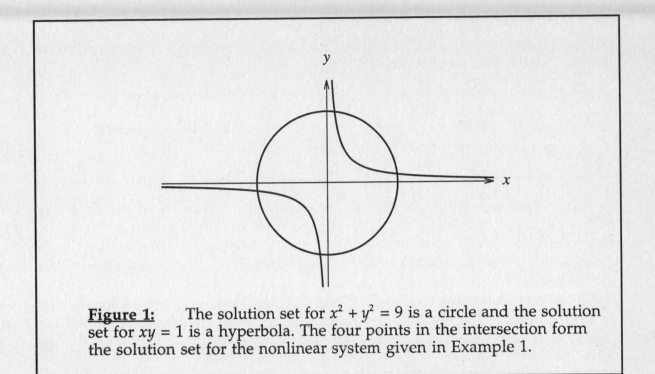

Figure 1: The solution set for $x^2 + y^2 = 9$ is a circle and the solution set for $xy = 1$ is a hyperbola. The four points in the intersection form the solution set for the nonlinear system given in Example 1.

The number of nonzero rows in reduced echelon form

In this subsection we show how the distinctive staircase-like shape of reduced echelon form imposes a limit on the number of nonzero rows. In particular, if an $(m \times n)$ matrix is in reduced echelon form, then the number of nonzero rows is *never* more than n.

We will use this result (stated in Theorem 1) to analyze the solution possibilities for a linear system. But, since the theorem is a little abstract, we introduce it with an example.

Example 2: Let B be a (5×2) nonzero matrix in reduced echelon form. Describe the possible forms B might take and verify that B can have no more than two nonzero rows.

Solution: The matrix B must have one of the following three patterns. Note that the number of nonzero rows is never greater than the number of columns in B, namely two.

$$\begin{bmatrix} 1 & 0 \\ 0 & 1 \\ 0 & 0 \\ 0 & 0 \\ 0 & 0 \end{bmatrix}, \quad \begin{bmatrix} 1 & \times \\ 0 & 0 \\ 0 & 0 \\ 0 & 0 \\ 0 & 0 \end{bmatrix}, \quad \begin{bmatrix} 0 & 1 \\ 0 & 0 \\ 0 & 0 \\ 0 & 0 \\ 0 & 0 \end{bmatrix}$$

∎

Example 2 is a special case of Theorem 1 which we state below. In words, Theorem 1 says that a matrix in reduced echelon form never has more nonzero rows than it has columns.

THEOREM 1   Let B be a $(p \times q)$ matrix in reduced echelon form. Let r denote the number of nonzero rows in B. Then $r \leq q$.

Proof: Without loss of generality, we can assume the leading 1 in the first row is in position $(1, 1)$. As we move down the rows of B, the leading 1 in each succeeding row must be at least one column to the right of the leading 1 in the row above. In other words, the leading 1 in each nonzero row is on or to the right of the main diagonal.

In particular, let the leading 1 in the last nonzero row be located in position (r, s). Since the leading 1 must be on or to the right of the main diagonal, it follows that $r \leq s$. Then, since $s \leq q$, we see that $r \leq q$. ∎

Solution possibilities for a linear system

The following theorem gives the precise answer to the question posed in the section title. Note that Theorem 2 is stated in terms of a *consistent* system. (If the system is inconsistent we know how many solutions there are—none).

THEOREM 2   Consider a consistent system of m linear equations in n unknowns. Let A denote the $(m \times (n + 1))$ augmented matrix for the system. Suppose, when A is transformed to reduced echelon form, that the resulting matrix B has exactly r nonzero rows.

Then $r \leq n$. Moreover:

(a)   If $r = n$, then the system has exactly one solution.
(b)   If $r < n$, then the system has infinitely many solutions, with $n - r$ independent variables.

Proof: We know from Theorem 1 that $r \leq n + 1$. Now, if $r = n + 1$, then the last nonzero row of B must have the form

$$[0, 0, \ldots, 0, 1].$$

But, if B has a row of the form above, then the original system must have been inconsistent. Since the original system was assumed consistent, the possibility $r = n + 1$ has been ruled out. Therefore, $r \leq n$.

Turning to case **(a)**, suppose $r = n$. As in the proof of Theorem 1, we can assume without loss of generality that the leading 1 in the first row of B is in the $(1, 1)$ position. In this case, the reduced matrix B has the following form:

$$B = \begin{bmatrix} 1 & 0 & 0 & \cdots & 0 & b_{1,n+1} \\ 0 & 1 & 0 & \cdots & 0 & b_{2,n+1} \\ 0 & 0 & 1 & \cdots & 0 & b_{3,n+1} \\ \vdots & & & & & \vdots \\ 0 & 0 & 0 & \cdots & 1 & b_{n,n+1} \\ 0 & 0 & 0 & \cdots & 0 & 0 \\ \vdots & & & & & \vdots \\ 0 & 0 & 0 & \cdots & 0 & 0 \end{bmatrix}.$$

Of course, if the reduced matrix B has the form displayed above, then the solution to the original system is $x_1 = b_{1,n+1}$, $x_2 = b_{2,n+2}$, . . . , $x_n = b_{n,n+1}$. This proves case **(a)**, showing that the original system has a unique solution.

Next, let us consider case **(b)**, $r < n$. In each nonzero row of B, the leading 1 in that row corresponds to a dependent variable. For example, if the leading 1 in row k is in position (k, j), then the kth equation is

$$x_j + b_{k,j+1} x_{j+1} + \cdots + b_{kn} x_n = b_{k,n+1} .$$

In the equation above, we can choose x_j as the dependent variable.

Since the number of nonzero rows is r, the number of dependent variables is also r. The remaining $n - r$ variables are independent variables and can take on any values whatsoever. Thus, the solution set is infinite. ■

For the sake of completeness we state the following corollary of Theorem 2.

COROLLARY   Given a system of linear equations, exactly one of the following possibilities holds:

(a)   The system has no solution.
(b)   The system has exactly one solution.
(c)   The system has infinitely many solutions.

The dimension of the solution set

Suppose a linear system has infinitely many solutions. We say (recall Section 3) that the solution set has ***dimension k*** if the general solution has k independent variables. Thus, see part **(b)** of Theorem 2, the solution set for a consistent linear system has dimension $n - r$ whenever $r < n$.

Example 3 Without solving the following system, we can say by the corollary that the system might have no solution, a single solution, or infinitely many solutions. If the system has infinitely many solutions, what are the possible values for the dimension of the solution set? Solve the system and confirm your answer.

$$x_1 + 2x_2 + 3x_3 = 1$$
$$4x_1 + 5x_2 + 6x_3 = 7$$
$$7x_1 + 8x_2 + 9x_3 = 13$$

Solution: For this system, $n = 3$. When the augmented matrix is transformed to reduced echelon form, it will be left with r nonzero rows, where r is either 1, 2, or 3.

If the system has infinitely many solutions, then we must have $r < n$ and therefore r is 1 or r is 2. So, we find that the solution set has either dimension $3 - 1 = 2$ or dimension $3 - 2 = 1$.

In fact, when the augmented matrix is transformed to reduced echelon form, we obtain:

$$\begin{bmatrix} 1 & 0 & -1 & 3 \\ 0 & 1 & 2 & -1 \\ 0 & 0 & 0 & 0 \end{bmatrix}.$$

From this, we see that the solution is $x_1 = 3 + x_3$, $x_2 = -1 - 2x_3$. In particular, the solution set has dimension 1. ∎

Exercises 1.7

In Exercises 1-6, determine all possibilities for the solution set (from among infinitely many solutions, a unique solution, or no solution) of the system of linear equations described.

1. A system of 3 equations in 4 unknowns.

2. A system of 2 equations in 3 unknowns that has $x_1 = -1$, $x_2 = 3$, $x_3 = -4$ as a solution.

3. A system of 4 equations in 4 unknowns.

4. A system of 4 equations in 4 unknowns that has $x_1 = 2$, $x_2 = 0$, $x_3 = -2$, $x_4 = 3$ as a solution.

5. A system of 4 equations in 3 unknowns.

6. A system of 4 equations in 3 unknowns that has $x_1 = 4$, $x_2 = -3$, $x_3 = 1$ as a solution.

In Exercises 7-10, the given system of equations is consistent. In each case, transform the augmented matrix to reduced echelon form and, in the notation of Theorem 2, determine n, r, and the number $n-r$, of independent variables. If $n - r > 0$, then identify $n - r$ independent variables.

7.
$$\begin{aligned} x_1 - 3x_2 + 2x_3 &= 0 \\ -2x_1 + 6x_2 - 3x_3 &= -1 \\ 3x_1 - 9x_2 + 8x_3 &= -2 \\ -x_1 + 3x_2 - 5x_3 &= 3 \end{aligned}$$

8.
$$\begin{aligned} x_1 + 3x_2 &= 10 \\ 2x_1 + 7x_2 &= 24 \\ x_1 + 4x_2 &= 14 \end{aligned}$$

9.
$$\begin{aligned} x_1 + 2x_2 + 5x_3 + 4x_4 &= -11 \\ -x_1 + x_3 - 2x_4 &= -1 \\ -x_1 + x_2 + 4x_3 - x_4 &= -7 \end{aligned}$$

10.
$$\begin{aligned} x_1 - 4x_2 + x_3 + 3x_4 &= 5 \\ 2x_1 - 8x_2 + 2x_3 + 7x_4 &= 12 \\ -x_1 + 4x_2 - x_3 - 5x_4 &= -9 \\ 3x_1 - 12x_2 + 3x_3 + 12x_4 &= 21 \end{aligned}$$

In Exercises 11 and 12, assume that the given system is consistent. For each system, determine, in the notation of Theorem 2, all possibilities for r and the number, $n - r$, of independent variables. Can the system have a unique solution?

11. $\begin{aligned} ax_1 + bx_2 &= c \\ dx_1 + ex_2 &= f \\ gx_1 + hx_2 &= i \end{aligned}$

12. $\begin{aligned} a_{11}x_1 + a_{12}x_2 + a_{13}x_3 + a_{14}x_4 &= b_1 \\ a_{21}x_1 + a_{22}x_2 + a_{23}x_3 + a_{24}x_4 &= b_2 \\ a_{31}x_1 + a_{32}x_2 + a_{33}x_3 + a_{34}x_4 &= b_3 \end{aligned}$

13. Let B be a (4×3) matrix in reduced echelon form.

(a) If B has three nonzero rows, exhibit all possible configurations for B.

(b) Suppose that a system of 4 linear equations in 2 unknowns has augmented matrix A, where A is a (4×3) matrix row equivalent to B. Demonstrate that the system of equations is inconsistent.

1.8 HOMOGENEOUS SYTEMS OF EQUATIONS

In the previous section we looked at the question of how many solutions a linear system might have. As we recall, a system might have no solution, one solution, or infinitely many solutions.

In this section we introduce a type of linear system, called a *homogeneous system*, where the possibility of no solution can be ruled out.

Homogeneous systems

A _homogeneous system of linear equations_ is a system where all the constants on the right-hand side are zero. For example, the following system of equations is a homogeneous system:

$$\begin{aligned}
x_1 + 2x_2 - x_3 &= 0 \\
2x_1 + 5x_2 - 4x_3 &= 0 \\
x_1 + 3x_3 &= 0.
\end{aligned}$$

As you can see by inspection, this homogeneous system must be consistent since one solution is given by $x_1 = 0$, $x_2 = 0$, $x_3 = 0$. In general, a homogeneous system is _always consistent_.

A homogeneous system is always consistent

Consider the general _(m × n) system of homogeneous equations_ :

(1)
$$\begin{aligned}
a_{11}x_1 + a_{12}x_2 + \cdots + a_{1n}x_n &= 0 \\
a_{21}x_1 + a_{22}x_2 + \cdots + a_{2n}x_n &= 0 \\
&\vdots \\
a_{m1}x_1 + a_{m2}x_2 + \cdots + a_{mn}x_n &= 0
\end{aligned}$$

(Again, the system (1) is called homogeneous because every constant on the right-hand side is zero.)

The homogeneous system (1) always has at least one solution, given by $x_1 = 0, x_2 = 0, \ldots, x_n = 0$. This solution is called the _zero solution_ or the _trivial solution._ All other solutions to (1), if it has any others, are called _nontrivial solutions_.

Example 1: Solve the system of equations. Does the system have nontrivial solutions?

$$\begin{aligned}
x_1 + 2x_2 - x_3 &= 0 \\
2x_1 + 5x_2 - 4x_3 &= 0 \\
x_1 + 3x_3 &= 0 .
\end{aligned}$$

<u>Solution:</u>   The augmented matrix can be reduced as follows:

$$\begin{bmatrix} 1 & 2 & -1 & 0 \\ 2 & 5 & -4 & 0 \\ 1 & 0 & 3 & 0 \end{bmatrix} \xrightarrow[R_3-R_1]{R_2-2R_1} \begin{bmatrix} 1 & 2 & -1 & 0 \\ 0 & 1 & -2 & 0 \\ 0 & -2 & 4 & 0 \end{bmatrix} \xrightarrow[R_3+2R_2]{R_1-2R_2} \begin{bmatrix} 1 & 0 & 3 & 0 \\ 0 & 1 & -2 & 0 \\ 0 & 0 & 0 & 0 \end{bmatrix}.$$

(Note that all the row operations preserve the column of zeros in the augmented matrix. Thus, there is no possibility of obtaining a row of the form $[0, 0, \ldots, 0, 1]$. This fact gives us another way of seeing that a homogeneous system is always consistent.)

Having transformed the augmented matrix into reduced echelon form, we see the original system is equivalent to

$$\begin{aligned} x_1 \quad + 3x_3 &= 0 \\ x_2 - 2x_3 &= 0. \end{aligned}$$

Therefore, the solution is $x_1 = -3x_3$, $x_2 = 2x_3$. Choosing x_3 to be nonzero will give us a nontrivial solution to the system. Choosing $x_3 = 0$ gives the trivial solution. ∎

When do homogeneous systems have nontrivial solutions?

As we have seen, a homogeneous system always has at least one solution; namely, the zero solution. In certain cases we can say more just by comparing the number of equations and the number of unknowns. The next theorem says that a homogeneous system *always* has nontrivial solutions whenever the system has more unknowns than equations.

THEOREM 1:   Consider the homogeneous system (1). If $m < n$, then the system has nontrivial solutions.

<u>Proof:</u>Because the system is homogeneous, it is consistent. Therefore we can apply Theorem 2 from Section 7. Let A denote the augmented matrix for the system (1) and let B denote the matrix that results when we transform A to reduced echelon form. Let r denote the number of nonzero rows in B.

Since B has a total of m rows, we know $r \le m$. Next, since $m < n$, we must conclude that $r < n$. Therefore, see part **(b)** of Theorem 2, the system has infinitely many solutions. (In fact, the solution set has dimension $n - r$.) ∎

Note that if $m \ge n$, then Theorem 1 gives us no information. In other words, when the number of equations, m, is greater than or equal to the number of unknowns, n, then we cannot know whether or not there are nontrivial solutions unless we solve the system to find out.

Example 2: What are the possibilities for the solution set of

$$x_1 + 2x_2 + x_3 + 3x_4 = 0$$
$$2x_1 + 4x_2 + 3x_3 + x_4 = 0$$
$$3x_1 + 6x_2 + 6x_3 + 2x_4 = 0 .$$

Solve the system.

Solution: Since the system is homogeneous and since it has more unknowns than equations, we see from Theorem 1 that there are nontrivial solutions.

The augmented matrix for the system transforms to the following reduced echelon form

$$\begin{bmatrix} 1 & 2 & 0 & 0 & 0 \\ 0 & 0 & 1 & 0 & 0 \\ 0 & 0 & 0 & 1 & 0 \end{bmatrix}.$$

Thus, the given system is equivalent to the following reduced system

$$x_1 + 2x_2 \qquad = 0$$
$$x_3 \qquad = 0$$
$$x_4 = 0 .$$

The general solution is $x_1 = -2x_2, x_3 = 0, x_4 = 0.$

Particular solutions can now be obtained by assigning values to the independent variable x_2. For instance, $x_1 = -2, x_2 = 1, x_3 = 0, x_4 = 0$ is one of the infinitely many nontrivial solutions to the system. ■

Example 3: What are the possibilities for the solution set of

$$2x_1 + 3x_2 - x_3 = 0$$
$$-2x_1 - 2x_2 + 3x_3 = 0$$
$$2x_1 + 6x_2 + 9x_3 = 0 .$$

Solve the system.

Solution: Theorem 1 does not apply to this system since $m = n = 3$. But, because the system is homogeneous, either the trivial solution is the unique solution or there are infinitely many solutions. To settle the issue, we have to solve the system. Transforming the augmented matrix to reduced echelon form, we obtain:

$$\begin{bmatrix} 1 & 0 & 0 & 0 \\ 0 & 1 & 0 & 0 \\ 0 & 0 & 1 & 0 \end{bmatrix}.$$

Thus, we find the trivial solution is the unique solution to the system.   ∎

Exercises 1.8

In Exercises 1-6 determine all possibilities for the solution set (from among infinitely many solutions, a unique solution, or no solution) of the system of linear equations described.

1. A homogeneous system of 3 equations in 3 unknowns.

2. A homogeneous system of 3 equations in 3 unknowns that has $x_1 = -1$, $x_2 = 3$, $x_3 = 4$ as a solution.

3. A homogeneous system of 4 equations in 3 unknowns.

4. A homogeneous system of 4 equations in 3 unknowns that has $x_1 = 2$, $x_2 = -3$, $x_3 = 6$ as a solution.

5. A homogeneous system of 3 equations in 4 unknowns.

6. A homogeneous system of 5 equations in 7 unknowns.

In Exercises 7-10, determine by inspection whether the given system has nontrivial solutions or only the trivial solution.

7.
$$\begin{aligned} 2x_1 + 3x_2 - x_3 &= 0 \\ x_1 - x_2 + 2x_3 &= 0 \end{aligned}$$

8.
$$\begin{aligned} x_1 + 2x_2 - x_3 + 2x_4 &= 0 \\ 2x_1 - 8x_2 + 2x_3 + 7x_4 &= 0 \\ 3x_1 - x_2 - 2x_3 + 3x_4 &= 0 \end{aligned}$$

9.
$$\begin{aligned} x_1 + 2x_2 - x_3 &= 0 \\ x_2 + 2x_3 &= 0 \\ 4x_3 &= 0 \end{aligned}$$

10.
$$\begin{aligned} x_1 - x_2 &= 0 \\ 3x_1 &= 0 \\ 2x_1 + x_2 &= 0 \end{aligned}$$

In Exercises 11-14, solve the given homogeneous system of equations by transforming the augmented matrix to reduced echelon form.

11.
$$\begin{aligned} x_1 + 3x_2 &= 0 \\ 2x_1 + 8x_2 &= 0 \\ 3x_1 + 6x_2 &= 0 \end{aligned}$$

12.
$$\begin{aligned} 4x_1 + 5x_2 + 8x_3 &= 0 \\ x_1 + x_2 + x_3 &= 0 \\ 3x_1 + 3x_2 + 7x_3 &= 0 \end{aligned}$$

$$
\begin{aligned}
13. \quad & 3x_1 + 5x_2 + 9x_3 = 0 \\
& 3x_1 + 4x_2 + 6x_3 = 0 \\
& 2x_1 + 3x_2 + 5x_3 = 0
\end{aligned}
$$

$$
\begin{aligned}
14. \quad & x_1 + 3x_2 + 3x_3 - 7x_4 + 10x_5 = 0 \\
& 2x_1 + 6x_2 + 4x_3 - 10x_4 + 14x_5 = 0 \\
& x_1 + 3x_2 \qquad\quad - x_4 + x_5 = 0
\end{aligned}
$$

15. Suppose $x_1 = s_1$, $x_2 = s_2$, $x_3 = s_3$ and $x_1 = t_1$, $x_2 = t_2$, $x_3 = t_3$ are solutions to the homogeneous system

$$
\begin{aligned}
ax_1 + bx_2 + cx_3 &= 0 \\
dx_1 + ex_2 + fx_3 &= 0
\end{aligned}
$$

(a) Demonstrate that $x_1 = s_1 + t_1$, $x_2 = s_2 + t_2$, $x_3 = s_3 + t_3$ is also a solution for the system.

(b) Demonstrate that $x_1 = ks_1$, $x_2 = ks_2$, $x_3 = ks_3$ is also a solution for the system, where k is any real number.

2.1 MATRIX ADDITION

In Chapter 1 we introduced matrices as a way to organize the calculations required to solve a system of linear equations. In this chapter we will see that matrices can be used in many other applications besides linear systems.

One reason matrices are so useful is that we can define arithmetic operations for matrices. In this section we show how to add matrices. Then, in Section 2, we show how to multiply matrices. We begin by introducing basic matrix notation.

Matrix Notation

We recall that an $(m \times n)$ matrix A is a rectangular array having m rows and n columns

$$A = \begin{bmatrix} a_{11} & a_{12} & \cdots & a_{1n} \\ a_{21} & a_{22} & \cdots & a_{2n} \\ \vdots & & & \vdots \\ a_{m1} & a_{m2} & \cdots & a_{mn} \end{bmatrix}.$$

The subscripts for the entry a_{ij} indicate that the entry appears in the ith row and the jth column. For example, a_{32} is the entry in row three, column two. We will frequently use the notation

$$A = (a_{ij})$$

to denote a matrix with entries a_{ij}.

Example 1   Display the (2×3) matrix $A = (a_{ij})$ where $a_{11} = 6$, $a_{12} = 3$, $a_{13} = 7$, $a_{21} = 2$, $a_{22} = 1$, and $a_{23} = 4$.

Solution:

$$A = \begin{bmatrix} 6 & 3 & 7 \\ 2 & 1 & 4 \end{bmatrix}.$$

■

Adding matrices

The definition of matrix addition is quite natural and we introduce it with an example. Consider the (3×2) matrices A and B:

$$A = \begin{bmatrix} 4 & -1 \\ 0 & 6 \\ 1 & 5 \end{bmatrix}, \quad B = \begin{bmatrix} 3 & 5 \\ 2 & -2 \\ 3 & 4 \end{bmatrix}.$$

If we want to define the sum, $A + B$, it is natural to add A and B by adding corresponding entries:

$$\begin{bmatrix} 4 & -1 \\ 0 & 6 \\ 1 & 5 \end{bmatrix} + \begin{bmatrix} 3 & 5 \\ 2 & -2 \\ 3 & 4 \end{bmatrix} = \begin{bmatrix} 7 & 4 \\ 2 & 4 \\ 4 & 9 \end{bmatrix}.$$

(In the example above, we calculated the $(1, 1)$ entry of $A + B$ by adding the corresponding $(1, 1)$ entries of A and B. Similarly, we found the $(1, 2)$ entry of $A + B$ by adding the $(1, 2)$ entries of A and B, and so forth.)

In general, let A and B be $(m \times n)$ matrices, $A = (a_{ij})$ and $B = (b_{ij})$. The **sum** $A + B$ is the $(m \times n)$ matrix C whose ijth entry is $a_{ij} + b_{ij}$. We can represent this definition symbolically as

$$(A + B)_{ij} = (A)_{ij} + (B)_{ij}.$$

Note that the definition of $A + B$ requires that A and B have the same number of rows and columns. For example, if A is (3×2) and B is (5×6), then the sum $A + B$ is not defined.

The following example illustrates matrix addition.

Example 2   Form $R + S$ and $R + T$ or state that the sum cannot be formed:

$$R = \begin{bmatrix} 3 & -2 \\ 1 & 6 \end{bmatrix}, \quad S = \begin{bmatrix} 6 & 5 \\ 3 & 2 \end{bmatrix}, \quad T = \begin{bmatrix} 4 & 6 & 1 \\ 0 & 3 & 3 \end{bmatrix}.$$

Solution: Since R is (2×2) and T is (2×3), the sum $R + T$ is not defined. The sum $R + S$ is defined and we find

$$R + S = \begin{bmatrix} 9 & 3 \\ 4 & 8 \end{bmatrix}$$

■

Scalar multiplication

Another useful operation is that of multiplying a matrix by a *scalar* (recall that "scalar" is just another name for "number"). For example, suppose we want to define the matrix $5A$ where

$$A = \begin{bmatrix} 3 & 1 \\ 6 & 0 \\ -2 & 4 \end{bmatrix}.$$

It is natural to define $5A$ to be the matrix that results when we multiply each entry of A by the scalar 5. In particular,

$$5A = \begin{bmatrix} 15 & 5 \\ 30 & 0 \\ -10 & 20 \end{bmatrix}.$$

In general, let A be an $(m \times n)$ matrix, $A = (a_{ij})$, and let s denote a scalar. The **_scalar multiple_** sA is the $(m \times n)$ matrix whose ijth entry is sa_{ij}.

Example 3   Form the scalar multiple $3R$ and the matrix $3R + 4S$

$$R = \begin{bmatrix} 1 & 3 & -1 \\ 0 & 6 & 2 \end{bmatrix}, \quad S = \begin{bmatrix} -2 & 1 & 6 \\ 2 & -1 & 5 \end{bmatrix}.$$

<u>Solution:</u>   The scalar multiple $3R$ is

$$3R = \begin{bmatrix} 3 & 9 & -3 \\ 0 & 18 & 6 \end{bmatrix}.$$

The combination $3R + 4S$ is given by

$$3R + 4S = \begin{bmatrix} 3 & 9 & -3 \\ 0 & 18 & 6 \end{bmatrix} + \begin{bmatrix} -8 & 4 & 24 \\ 8 & -4 & 20 \end{bmatrix} = \begin{bmatrix} -5 & 13 & 21 \\ 8 & 14 & 26 \end{bmatrix}.$$

∎

Algebraic properties of matrix addition and scalar multiplication

Whenever we design new arithmetic operations, such as matrix addition and scalar multiplication, we naturally want to investigate the algebraic properties of these new operations. As the theorem below says, these operations hold no surprises.

THEOREM 1   Let A, B, and C be $(m \times n)$ matrices and let r and s be scalars. Then

(a) $A + B = B + A$

(b) $(A + B) + C = A + (B + C)$

(c) $(r + s)A = rA + sA$

(d) $r(A + B) = rA + rB$

(e) $(rs)A = r(sA)$

The proof of Theorem 1 is a relatively straightforward application of familiar algebraic properties from the real numbers. We omit the proof.

In the next section we describe how to multiply two matrices. As we see in that section, matrix multiplication does not behave as nicely as matrix addition and scalar multiplication. For instance, we will see that AB is not generally equal to BA. We will also see examples where $AB = 0$, but neither A nor B is 0.

Exercises 2.1

1. Display the (2x3) matrix $A = (a_{ij})$, where $a_{11} = 2$, $a_{12} = 1$, $a_{13} = 6$, $a_{21} = 4$, $a_{22} = 3$, $a_{23} = 8$.

2. Display the (2x4) matrix $C = (c_{ij})$, where $c_{23} = 4$, $c_{12} = 2$, $c_{21} = 2$, $c_{14} = 1$, $c_{22} = 2$, $c_{24} = 3$, $c_{11} = 1$, $c_{13} = 7$.

3. Display the (3x3) matrix $Q = (q_{ij})$, where $q_{23} = 1$, $q_{32} = 2$, $q_{11} = 1$, $q_{13} = -3$, $q_{22} = 1$, $q_{33} = 1$, $q_{21} = 2$, $q_{12} = 4$, $q_{31} = 3$.

4. If $\begin{bmatrix} v + 2w & x - y \\ 2x + y & -v + w \end{bmatrix} = \begin{bmatrix} 1 & -6 \\ 0 & -4 \end{bmatrix}$

   then determine the values of v, w, x, and y.

Use the following (2x2) matrices in Exercises 5-10.

$$A = \begin{bmatrix} 2 & 1 \\ 1 & 3 \end{bmatrix} \quad B = \begin{bmatrix} 0 & -1 \\ 1 & 3 \end{bmatrix} \quad C = \begin{bmatrix} -2 & 3 \\ 1 & 1 \end{bmatrix} \quad O = \begin{bmatrix} 0 & 0 \\ 0 & 0 \end{bmatrix}$$

5. Find: (a) $A + B$; (b) $A + C$; (c) $6B$; (d) $B + 3C$.

6. Find: (a) $B + C$; (b) $3A$; (c) $A + 2C$; (d) $C + 8O$.

7. Find a matrix D such that $A + D = B$.

8. Find a matrix D such that $A + 2D = C$.

9. Find a matrix D such that $A + 2B + 2D = 3B$.

10. Find a matrix D such that $2A + 5B + D = 2B + 3A$.

11. If $D = \begin{bmatrix} 6 & 7 \\ -1 & -3 \end{bmatrix}$ then find scalars x_1 and x_2 such that $x_1 A + x_2 B = D$.

12. If $D = \begin{bmatrix} -8 & 6 \\ 4 & 6 \end{bmatrix}$ then find scalars x_1, x_2, and x_3 such that $x_1 A + x_2 B + x_3 C = D$.

2.2 MATRIX MULTIPLICATION

In the previous section we defined two matrix operations, addition and scalar multiplication. These operations are natural extensions of addition and multiplication for real numbers. Such is not the case with the operation we define in this section—matrix multiplication.

Although the definition of matrix multiplication is complicated, it arises naturally in practical applications of matrices.

Multiplying row matrices times column matrices

The basic calculation involved in multiplying two matrices is that of multiplying a row matrix times a column matrix. To illustrate a row-column product, consider

$$\begin{bmatrix} -3 & 2 & 4 \end{bmatrix} \begin{bmatrix} 2 \\ -1 \\ 3 \end{bmatrix} = (-3)(2) + (2)(-1) + (4)(3) = 4.$$

In general, let **a** be a row matrix and let **b** be a column matrix:

$$\mathbf{a} = \begin{bmatrix} a_1 & a_2 & \cdots & a_n \end{bmatrix} \quad \text{and} \quad \mathbf{b} = \begin{bmatrix} b_1 \\ b_2 \\ \vdots \\ b_n \end{bmatrix}.$$

The product **ab** is defined by

$$\mathbf{ab} = \begin{bmatrix} a_1 & a_2 & \cdots & a_n \end{bmatrix} \begin{bmatrix} b_1 \\ b_2 \\ \vdots \\ b_n \end{bmatrix} = a_1 b_1 + a_2 b_2 + \cdots + a_n b_n.$$

In order for the product **ab** to be defined, it is necessary that **a** and **b** each have the same number of components. Thus, if **a** is a $(1 \times n)$ row matrix then **b** must be an $(n \times 1)$ column matrix. For example, the following products are not defined:

$$\begin{bmatrix} 2 & 1 & -1 \end{bmatrix} \begin{bmatrix} 3 \\ 2 \end{bmatrix} \quad \text{and} \quad \begin{bmatrix} 2 & 5 \end{bmatrix} \begin{bmatrix} 3 \\ 2 \\ 4 \end{bmatrix}.$$

Observe that, when defined, the product **ab** is a scalar.

An example of matrix multiplication

The product AB of two matrices is determined by multiplying the rows of A times the columns of B.

We introduce matrix multiplication by way of an example. Consider the two matrices:

$$A = \begin{bmatrix} 3 & 1 & -4 \\ 2 & -2 & 3 \end{bmatrix} \ , \quad B = \begin{bmatrix} -1 & 1 \\ 3 & -2 \\ 1 & 4 \end{bmatrix} .$$

Since each row of A has three entries and each column of B also has three entries, we can multiply any row of A times any column of B. In fact, since A has two different rows and B has two different columns, we can form four different row-column product combinations:

$$AB = \begin{bmatrix} 3 & 1 & -4 \\ 2 & -2 & 3 \end{bmatrix}\begin{bmatrix} -1 & 1 \\ 3 & -2 \\ 1 & 4 \end{bmatrix} = \begin{bmatrix} c_{11} & c_{12} \\ c_{21} & c_{22} \end{bmatrix}$$

In particular, the entry c_{11} is formed by multiplying row 1 of A times column 1 of B:

$$c_{11} = \begin{bmatrix} 3 & 1 & -4 \end{bmatrix}\begin{bmatrix} -1 \\ 3 \\ 1 \end{bmatrix} = (3)(-1) + (1)(3) + (-4)(1) = -4.$$

Similarly, c_{12} is the product of row 1 of A times column 2 of B, c_{21} is the product of row 2 of A times column 1 of B, and c_{22} is the product of row 2 of A times column 2 of B. In summary, we have

$$AB = \begin{bmatrix} 3 & 1 & -4 \\ 2 & -2 & 3 \end{bmatrix}\begin{bmatrix} -1 & 1 \\ 3 & -2 \\ 1 & 4 \end{bmatrix} = \begin{bmatrix} -4 & -15 \\ -5 & 18 \end{bmatrix} .$$

Having seen the example above, we are ready to give the formal definition of matrix multiplication.

The definition of matrix multiplication

Suppose A and B are matrices where the rows of A can be multiplied times the columns of B. Then, as in the above example, we can define the product AB. Also, as in the example, the product AB will have the same number of rows as A and the same number of columns as B.

DEFINITION 1   Let $A = (a_{ij})$ be an $(m \times n)$ matrix and let $B = (b_{ij})$ be an $(r \times s)$ matrix. If $n = r$, then the ***product*** AB is the $(m \times s)$ matrix whose ij-th entry is defined by

$$(AB)_{ij} = a_{i1}b_{1j} + a_{i2}b_{2j} + \cdots + a_{in}b_{nj}$$

If $n \neq r$, then the product AB is not defined.

The definition can be visualized as in Figure 1.

$$
\begin{bmatrix}
a_{11} & a_{12} & \cdots & a_{1n} \\
\vdots & & & \vdots \\
a_{i1} & a_{i2} & \cdots & a_{in} \\
\vdots & & & \vdots \\
a_{m1} & a_{m2} & \cdots & a_{mn}
\end{bmatrix}
\begin{bmatrix}
b_{11} & \cdots & b_{1j} & \cdots & b_{1s} \\
b_{21} & & b_{2j} & & b_{2s} \\
\vdots & & \vdots & & \vdots \\
b_{n1} & \cdots & b_{nj} & \cdots & b_{ns}
\end{bmatrix}
=
\begin{bmatrix}
c_{11} & \cdots & c_{1j} & \cdots & c_{1s} \\
\vdots & & \vdots & & \vdots \\
c_{i1} & \cdots & [c_{ij}] & \cdots & c_{is} \\
\vdots & & \vdots & & \vdots \\
c_{m1} & & c_{mj} & & c_{ms}
\end{bmatrix}
$$

Figure 1 c_{ij} is the ij-th entry of AB. c_{ij} is the product of row i of A and column j of B.

Example 1   Let the matrices A, B, C, and D be given by

$$
A = \begin{bmatrix} 1 & 2 \\ 2 & 3 \end{bmatrix}, \quad
B = \begin{bmatrix} -3 & 2 \\ 1 & -2 \end{bmatrix}, \quad
C = \begin{bmatrix} 1 & 0 & -2 \\ 0 & 1 & 1 \end{bmatrix}, \quad
D = \begin{bmatrix} 3 & 1 \\ -1 & -2 \\ 1 & 1 \end{bmatrix}.
$$

Find each of AB, BA, AC, CA, CD, and DC or state that the indicated product cannot be formed.

Solution: The definition of matrix multiplication yields

$$AB = \begin{bmatrix} -1 & -2 \\ -3 & -2 \end{bmatrix} \ , \quad BA = \begin{bmatrix} 1 & 0 \\ -3 & -4 \end{bmatrix} \ , \quad AC = \begin{bmatrix} 1 & 2 & 0 \\ 2 & 3 & -1 \end{bmatrix}.$$

The product CA is undefined. The products CD and DC are given by

$$CD = \begin{bmatrix} 1 & -1 \\ 0 & -1 \end{bmatrix} \ , \quad DC = \begin{bmatrix} 3 & 1 & -5 \\ -1 & -2 & 0 \\ 1 & 1 & -1 \end{bmatrix}.$$

∎

Visualizing the size of the product

Let A be $(m \times n)$ and let B be $(n \times s)$ so that we can form the product AB. To visualize the size of the product it is helpful to copy down the dimensions as in Figure 1:

$$(m \times n)\,(n \times s) = (m \times s)$$

In particular, the product is not defined unless the inner dimensions agree. If the product is defined, then the outer dimensions give the size of the product.

Example 2 Let A and B be matrices of the size given. Determine whether or not the product can be formed. If it can, give the size of the product.

(a) AB and BA where A is (2×4) and B is (4×2)
(b) AB and BA where A is (4×4) and B is (2×4)
(c) AB and BA where A is (3×5) and B is (3×5)

Solution: (a) AB is defined and the product is $(2 \times 4)\,(4 \times 2) = (2 \times 2)$. Similarly, BA is defined and is (4×4).

(b) AB is not defined, but BA is. The size of BA is $(2 \times 4)\,(4 \times 4) = (2 \times 4)$.

(c) Neither AB nor BA is defined.

∎

Algebraic properties of matrix multiplication

Example 1 illustrates that matrix multiplication does not behave exactly like real number multiplication. In particular, if a and b are real numbers, then we know that

$$ab = ba.$$

But (see Example 1), when A and B are matrices we often have

$$AB \neq BA.$$

The next theorem lists some properties of matrix multiplication that do hold.

THEOREM 1   The following associative and distributive properties hold:

(a)   Let A, B, and C be $(m \times n)$, $(n \times p)$, and $(p \times q)$ respectively. Then $(AB)C = A(BC)$.

(b)   Let A and B be $(m \times n)$ and $(n \times p)$ respectively and let r be a scalar. Then $r(AB) = (rA)B = A(rB)$.

(c)   Let A and B be $(m \times n)$ and let C be $(n \times p)$. Then $(A + B)C = AC + BC$.

(d)   Let A be $(m \times n)$ and let B and C be $(n \times p)$. Then $A(B + C) = AB + AC$.

Exercises 2.2

In Exercises 1-9 use the matrices

$$A = \begin{bmatrix} 2 & 1 \\ 1 & 3 \end{bmatrix} \quad B = \begin{bmatrix} 0 & -1 \\ 1 & 3 \end{bmatrix} \quad C = \begin{bmatrix} -2 & 3 \\ 1 & 1 \end{bmatrix} \quad O = \begin{bmatrix} 0 & 0 \\ 0 & 0 \end{bmatrix}$$

$$D = \begin{bmatrix} 2 & 1 \\ 4 & 0 \\ 8 & -1 \\ 3 & 2 \end{bmatrix} \quad E = \begin{bmatrix} 2 & 1 & 3 & 6 \\ 2 & 0 & 0 & 4 \\ 1 & -1 & 1 & -1 \\ 1 & 3 & 1 & 2 \end{bmatrix}$$

to perform the indicated operations, or state that the operation is not defined.

1. $(A+B)C$ and $AC + BC$ 2. AB and BA 3. AO and OA

4. $(A+O)B$ 5. DE and ED 6. AD and DA

7. AA 8. DD 9. EE

10. Determine whether the following matrix products are defined. When the product is defined, give the size of the product.

     (a) AB and BA, where A is (2×3) and B is (3×4).

     (b) AB and BA, where A is (2×3) and B is (2×4).

     (c) AB and BA, where A is (3×7) and B is (6×3).

     (d) AB and BA, where A is (2×3) and B is (3×2).

     (e) AB and BA, where A is (3×3) and B is (3×1).

     (f) $A(BC)$ and $(AB)C$, where A is (2×3), B is (3×5), and C is (5×4).

     (g) AB and BA, where A is (4×1) and B is (1×4).

11. What is the size of the product $(AB)(CD)$, where A is (2×3), B is (3×4), C is (4×4), and D is (4×2)? Also, calculate the size of

$$A[B(CD)] \text{ and } [(AB)C]D.$$

12. Let $A = \begin{bmatrix} 1 & 2 \\ 3 & 4 \end{bmatrix}$.

(a) Find all matrices $B = \begin{bmatrix} a & b \\ c & d \end{bmatrix}$ such that $AB = BA$.

(b) Use the results of part (a) to exhibit (2×2) matrices B and C such that $AB = BA$ and $AC \neq CA$.

13. Let A and B be matrices such that the product AB is defined and is a square matrix. Argue that the product BA is also defined and is a square matrix.

2.3 THE IDENTITY MATRIX, THE ZERO MATRIX, MATRIX POWERS

This section will introduce two special matrices, the *identity matrix* and the *zero matrix*. We also define powers of a square matrix and look further at how the rules for matrix multiplication differ from the familiar rules for real number multiplication.

The identity matrix

The real number 1 is called the *multiplicative identity* for the real numbers because it has the property that

$$1a = a1 = a$$

for any real number a.

Similarly, for each positive integer n, there is an $(n \times n)$ matrix denoted I_n that has the property that

$$I_n A = A I_n = A$$

for any $(n \times n)$ matrix A. The matrix I_n is called the **$(n \times n)$ identity matrix** and is defined by

$$I_n = \begin{bmatrix} 1 & 0 & 0 & \cdots & 0 \\ 0 & 1 & 0 & \cdots & 0 \\ 0 & 0 & 1 & \cdots & 0 \\ \vdots & & & & \vdots \\ 0 & 0 & 0 & \cdots & 1 \end{bmatrix}.$$

Formally, the ij-th entry of I_n is 0 when $i \neq j$ and is 1 when $i = j$. For example, the (2×2) and (3×3) identity matrices are given by

$$I_2 = \begin{bmatrix} 1 & 0 \\ 0 & 1 \end{bmatrix}, I_3 = \begin{bmatrix} 1 & 0 & 0 \\ 0 & 1 & 0 \\ 0 & 0 & 1 \end{bmatrix}.$$

The following theorem shows that I_n plays the same role for matrix multiplication that the number 1 plays for real number multiplication.

THEOREM 1   Let B be an ($m \times n$) matrix. Then

    (a) $I_m B = B$
    (b) $BI_n = B$.

In particular, if A is an ($n \times n$) matrix, then $I_n A = AI_n = A$.

Note from Theorem 1 that the peculiarities of matrix multiplication force us to use different size identity matrices when B is not a square matrix. For example, if B is a (5×3) matrix, then we have:

$$I_5 B = B \quad \text{and} \quad BI_3 = B.$$

Usually the dimension of the identity matrix is clear from the context of the problem. In that case it is customary to drop the subscript n and denote the ($n \times n$) identity matrix simply as I. For example, if A is a (5×5) matrix, then we will write $IA = AI = A$ rather than $I_5 A = AI_5 = A$.

The zero matrix

The real number 0 is called the *additive identity* for the real numbers because it has the property that

$$0 + a = a + 0 = a$$

for any real number a. Similarly, we can define a zero matrix that serves as an identity for matrix addition.

In particular, we define the **_($m \times n$) zero matrix_**, to be the ($m \times n$) matrix O_{mn} that consists entirely of zeros. For example, the matrix O_{32} below is the (3×2) zero matrix:

$$O_{32} = \begin{bmatrix} 0 & 0 \\ 0 & 0 \\ 0 & 0 \end{bmatrix}.$$

As with the identity matrix, we will drop the subscripts and simply denote the zero matrix as O when the dimension of O is clear from the context of the problem we are working. It is easy to see that the zero matrix is an identity for matrix addition; that is,

$$O + A = A + O = A.$$

Matrix powers

We are familiar with using exponent notation as shorthand way of writing powers of a real number:

$$aa = a^2, \quad aaa = a^3, \quad \text{etc.}$$

Similarly, if A is an $(n \times n)$ matrix, then we define

$$A^2 = AA, \quad A^3 = AAA, \quad \text{etc.}$$

In general, if k is a *positive integer* and A is a *square* matrix, we define the **_power_** A^k to be the matrix formed by multiplying A by itself k times. We also define the special power A^0 by $A^0 = I$.

Finally, note that if B is an $(m \times n)$ matrix, then we *cannot* define B^2 unless $m = n$. In general, B^k only makes sense when B is a square matrix.

Irregular aspects of matrix multiplication

We cannot use matrix multiplication effectively unless we are comfortable with the rules. We listed some of these rules in Theorem 1 of the previous section.

If we compare the rules for matrix multiplication with the familiar rules for multiplying real numbers, we see some notable differences. For example, we can always multiply any two real numbers, but:

We cannot form the matrix product AB unless the number of columns in A is the same as the number of rows of B.

For real numbers, the order of the product is not important, we always have $ab = ba$. For matrices, even if we can form both of the products AB and BA:

We usually find that AB ≠ BA.

The next two examples point out some other irregular aspects of matrix multiplication, namely:

(a)   _AB = O does not imply that either A or B is the zero matrix._
(b)   _AB = AC does not imply that B = C._

Example 1   For real numbers: If $ab = 0$, then either $a = 0$ or $b = 0$. Find a pair of (2×2) matrices A and B such that $AB = O$ but neither A nor B is the zero matrix.

Solution: There are many such pairs A and B. One example is the following:

$$A = \begin{bmatrix} 2 & -2 \\ -1 & 1 \end{bmatrix}, B = \begin{bmatrix} 3 & 4 \\ 3 & 4 \end{bmatrix}.$$

When we form the product AB, we get the zero matrix O:

$$\begin{bmatrix} 2 & -2 \\ -1 & 1 \end{bmatrix}\begin{bmatrix} 3 & 4 \\ 3 & 4 \end{bmatrix} = \begin{bmatrix} 0 & 0 \\ 0 & 0 \end{bmatrix}.$$

So, even though $AB = O$, we cannot necessarily assume that either $A = O$ or $B = O$. ∎

Example 2 For real numbers: If $ab = ac$ and if $a \neq 0$, then $b = c$. Find (2×2) matrices A, B, and C such that $AB = AC$ but $B \neq C$.

Solution: There are many different choices of A, B, and C that satisfy this property. For example, consider

$$A = \begin{bmatrix} 2 & -2 \\ -1 & 1 \end{bmatrix}, B = \begin{bmatrix} 5 & 6 \\ 2 & 7 \end{bmatrix}, C = \begin{bmatrix} 2 & 2 \\ -1 & 3 \end{bmatrix}.$$

As is easy to check,

$$AB = \begin{bmatrix} 2 & -2 \\ -1 & 1 \end{bmatrix}\begin{bmatrix} 5 & 6 \\ 2 & 7 \end{bmatrix} = \begin{bmatrix} 6 & -2 \\ -3 & 1 \end{bmatrix}$$

$$AC = \begin{bmatrix} 2 & -2 \\ -1 & 1 \end{bmatrix}\begin{bmatrix} 2 & 2 \\ -1 & 3 \end{bmatrix} = \begin{bmatrix} 6 & -2 \\ -3 & 1 \end{bmatrix}$$

So, even though $AB = AC$, we cannot necessarily assume that $B = C$. ∎

Exercises 2.3

1. In (a)-(c), determine n and m so that $I_n A = A$ and $A I_m = A$.

   (a) A is (2x3)　　　(b) A is (5x7)　　　(c) A is (4x4)

In Exercises 2-12, use the matrices

$$A = \begin{bmatrix} -1 & 3 \\ 2 & 1 \end{bmatrix}, \quad B = \begin{bmatrix} 0 & -1 & 4 \\ 3 & 2 & -3 \end{bmatrix}, \quad C = \begin{bmatrix} 1 & -1 \\ 3 & -4 \\ 0 & 2 \end{bmatrix}.$$

to perform the indicated operations, or state that the operation is not defined.

2. $A + O_{22}$ 　　　3. $B + O_{22}$ 　　　4. $C + O_{32}$

5. $A O_{23}$ 　　　6. $C O_{22}$ 　　　7. $B O_{23}$

8. $(AB) O_{32}$ 　　　9. $O_{22}(BC)$ 　　　10. A^3

11. B^2 　　　12. $C A^2$

13. Calculate A^3, where $A = \begin{bmatrix} 0 & 1 & -1 \\ 0 & 0 & 3 \\ 0 & 0 & 0 \end{bmatrix}$.

14. Let $A = \begin{bmatrix} 1 & -1 \\ 0 & 1 \end{bmatrix}$ and $B = \begin{bmatrix} 2 & 1 \\ 3 & 0 \end{bmatrix}$.

    Calculate $A^2 - B^2$ and $(A - B)(A + B)$. Explain why $A^2 - B^2 \neq (A - B)(A + B)$.

15. With A and B as in Exercise 14, calculate $(A+B)^2$ and $A^2 + 2AB + B^2$. Explain why $(A + B)^2 \neq A^2 + 2AB + B^2$.

16. Let $B = \begin{bmatrix} 1 & -3 \\ 2 & -6 \end{bmatrix}$. Describe all matrices $A = \begin{bmatrix} a & b \\ c & d \end{bmatrix}$ such that $AB = O_{22}$.

70

17. Let $B = \begin{bmatrix} 4 & -1 \\ -2 & -3 \end{bmatrix}$ and $C = \begin{bmatrix} 3 & 2 \\ -4 & 3 \end{bmatrix}$. Describe all matrices $A = \begin{bmatrix} a & b \\ c & d \end{bmatrix}$ such that $AB = AC$.

18. Set $A = \begin{bmatrix} 2 & 0 \\ 0 & 2 \end{bmatrix}$ and $B = \begin{bmatrix} 1 & b \\ b^{-1} & 1 \end{bmatrix}$, where $b \neq 0$. Show that O_{22}, A, and B are solutions to the matrix equation $X^2 - 2X = O_{22}$. Conclude that the "quadratic equation" $X^2 - 2X = O_{22}$ has infinitely many matrix solutions.

2.4 VECTORS IN R^n; THE EQUATION $A\mathbf{x} = \mathbf{b}$

In Chapter 1 we used the augmented matrix to represent a system of linear equations. In this section we see how a system can be represented as the simple matrix equation $A\mathbf{x} = \mathbf{b}$.

Vectors

In science and engineering, a quantity having both magnitude and direction is referred to as a *vector*. Familiar examples of vector quantities include force vectors and velocity vectors.

In matrix theory, we use the term *vector* to denote a matrix having just one row or just one column (in a later chapter we will see how physical vectors such as velocity can be represented in terms of the algebraic vectors we discuss in this section).

We will refer to a vector as being *n-dimensional* if it has n components. For instance, the vector $\mathbf{x}$ is an example of a five-dimensional row vector:

$$\mathbf{x} = [3, 7, 0, -5, 2].$$

Similarly, $\mathbf{y}$ is an example of a four-dimensional column vector:

$$\mathbf{y} = \begin{bmatrix} 2 \\ 1 \\ 8 \\ 3 \end{bmatrix}.$$

Throughout this text, unless we state otherwise, the term vector will always refer to a column vector. We will denote vectors with boldface type.

The set of *n*-dimensional vectors, R^n

It is common to denote the set of all n-dimensional vectors by the symbol R^n. Stated formally, R^n is the set defined as follows:

$$R^n = \left\{ \mathbf{x} : \mathbf{x} = \begin{bmatrix} x_1 \\ x_2 \\ \vdots \\ x_n \end{bmatrix}, \text{where } x_1, x_2, \ldots, x_n \text{ are real numbers} \right\}.$$

Since a vector in R^n can be regarded as an $(n \times 1)$ matrix, addition and scalar multiplication are defined for vectors. For example, the vectors $\mathbf{u}$ and $\mathbf{v}$ are in R^3, as is the sum $\mathbf{u} + \mathbf{v}$:

$$\mathbf{u} = \begin{bmatrix} 2 \\ -3 \\ 5 \end{bmatrix} \quad , \quad \mathbf{v} = \begin{bmatrix} 1 \\ 5 \\ 2 \end{bmatrix} \quad , \quad \mathbf{u} + \mathbf{v} = \begin{bmatrix} 3 \\ 2 \\ 7 \end{bmatrix} .$$

Matrix representation for a system of linear equations

The definition of matrix multiplication leads to a shorthand notation for a system of linear equations. For example, consider the linear system

$$x_1 + 2x_2 + 7x_3 = 2$$
$$x_1 + 5x_2 + 2x_3 = 3$$

We can rewrite the system above as a single vector equation:

$$\begin{bmatrix} x_1 + 2x_2 + 7x_3 \\ x_1 + 5x_2 + 2x_3 \end{bmatrix} = \begin{bmatrix} 2 \\ 3 \end{bmatrix} .$$

Next, using the definition of matrix multiplication, we can rewrite the single vector equation as follows:

$$\begin{bmatrix} 1 & 2 & 7 \\ 1 & 5 & 2 \end{bmatrix} \begin{bmatrix} x_1 \\ x_2 \\ x_3 \end{bmatrix} = \begin{bmatrix} 2 \\ 3 \end{bmatrix} .$$

Finally, the equation above can be written simply as

$$A\mathbf{x} = \mathbf{b}$$

where A, $\mathbf{x}$, and $\mathbf{b}$ are:

$$A = \begin{bmatrix} 1 & 2 & 7 \\ 1 & 5 & 2 \end{bmatrix} \quad , \quad \mathbf{x} = \begin{bmatrix} x_1 \\ x_2 \\ x_3 \end{bmatrix} \quad , \quad \mathbf{b} = \begin{bmatrix} 2 \\ 3 \end{bmatrix} .$$

The example above suggests that we can rewrite any $(m \times n)$ system of linear equations in the form $A\mathbf{x} = \mathbf{b}$.

In particular, consider an $(m \times n)$ system of linear equations of the form

$$
\begin{aligned}
a_{11}x_1 + a_{12}x_2 + \cdots + a_{1n}x_n &= b_1 \\
a_{21}x_1 + a_{22}x_2 + \cdots + a_{2n}x_n &= b_2 \\
&\ \ \vdots \\
a_{m1}x_1 + a_{m2}x_2 + \cdots + a_{mn}x_n &= b_m
\end{aligned}
$$

(1)

The system (1) can be rewritten as the matrix equation

(2) $$Ax = b$$

where

$$
A = \begin{bmatrix} a_{11} & a_{12} & \cdots & a_{1n} \\ a_{21} & a_{22} & & a_{2n} \\ \vdots & & & \vdots \\ a_{m1} & a_{m2} & \cdots & a_{mn} \end{bmatrix} , \quad
x = \begin{bmatrix} x_1 \\ x_2 \\ \vdots \\ x_n \end{bmatrix} , \quad
b = \begin{bmatrix} b_1 \\ b_2 \\ \vdots \\ b_m \end{bmatrix} .
$$

The matrix A is called the _**coefficient matrix**_ for the system (1).

Example 1: Consider the system of equations

$$
\begin{aligned}
2x_1 + \ x_2 + \ 4x_3 &= 7 \\
x_1 + \ x_2 + \ 3x_3 &= 2 \\
4x_1 + 6x_2 + 17x_3 &= 5
\end{aligned}
$$

(a) Suppose we want to rewrite this system as $Ax = b$. Identify the coefficient matrix A and the vectors x and b.

(b) Verify that the vector s is a solution to the system, where

$$
s = \begin{bmatrix} s_1 \\ s_2 \\ s_3 \end{bmatrix} = \begin{bmatrix} 2 \\ -9 \\ 3 \end{bmatrix} .
$$

Solution: **(a)** The coefficient matrix A and the vectors x and b are:

$$
A = \begin{bmatrix} 2 & 1 & 4 \\ 1 & 1 & 3 \\ 4 & 6 & 17 \end{bmatrix} , \quad
x = \begin{bmatrix} x_1 \\ x_2 \\ x_3 \end{bmatrix} , \quad
b = \begin{bmatrix} 7 \\ 2 \\ 5 \end{bmatrix} .
$$

(b) A direct multiplication shows that $A\mathbf{s} = \mathbf{b}$:

$$A\mathbf{s} = \begin{bmatrix} 2 & 1 & 4 \\ 1 & 1 & 3 \\ 4 & 6 & 17 \end{bmatrix} \begin{bmatrix} 2 \\ -9 \\ 3 \end{bmatrix} = \begin{bmatrix} 7 \\ 2 \\ 5 \end{bmatrix} = \mathbf{b}.$$

∎

Notation for an augmented matrix

As we have seen, we can use the shorthand notation $A\mathbf{x} = \mathbf{b}$ to denote a system of linear equations. If we want to *solve* this system of equations, however, we have to form the augmented matrix and then transform the augmented matrix to reduced echelon form.

It is convenient, therefore, to have a standard notation for an augmented matrix. We use the symbol

$$[\,A \mid \mathbf{b}\,]$$

to denote the augmented matrix for the system $A\mathbf{x} = \mathbf{b}$. We will illustrate this notation in the next subsection.

The vector form for the general solution

As a final type of simplifying notation, we derive a compact way to express the solution of a linear system. We call this expression the _**vector form for the general solution**_.

The idea of the vector form for the general solution is best explained by some examples.

Example 2   The matrix $[\,A \mid \mathbf{0}\,]$ is the augmented matrix for a homogeneous system of linear equations:

$$[\,A \mid \mathbf{0}\,] = \begin{bmatrix} 1 & 0 & -1 & -3 & 0 \\ 0 & 1 & 2 & 1 & 0 \end{bmatrix}.$$

Find the general solution and express it in terms of vectors.

Solution: Since $[\,A \mid \mathbf{0}\,]$ is in reduced echelon form, it is easy to write the general solution:

$$x_1 = x_3 + 3x_4 \qquad \text{and} \qquad x_2 = -2x_3 - x_4 \,.$$

In vector form, therefore, the general solution can be expressed as

$$\mathbf{x} = \begin{bmatrix} x_1 \\ x_2 \\ x_3 \\ x_4 \end{bmatrix} = \begin{bmatrix} x_3 + 3x_4 \\ -2x_3 - x_4 \\ x_3 \\ x_4 \end{bmatrix} = \begin{bmatrix} x_3 \\ -2x_3 \\ x_3 \\ 0 \end{bmatrix} + \begin{bmatrix} 3x_4 \\ -x_4 \\ 0 \\ x_4 \end{bmatrix} = x_3 \begin{bmatrix} 1 \\ -2 \\ 1 \\ 0 \end{bmatrix} + x_4 \begin{bmatrix} 3 \\ -1 \\ 0 \\ 1 \end{bmatrix}.$$

The last expression above is what we call the *vector form for the general solution.* ∎

We comment briefly on Example 2. Even though there are infinitely many solutions to the system in Example 2, we have expressed them all in vector form as

$$\mathbf{x} = a\mathbf{v}_1 + b\mathbf{v}_2$$

where a and b are arbitrary scalars and where $\mathbf{v}_1$ and $\mathbf{v}_2$ are the vectors

$$\mathbf{v}_1 = \begin{bmatrix} 1 \\ -2 \\ 1 \\ 0 \end{bmatrix}, \quad \mathbf{v}_2 = \begin{bmatrix} 3 \\ -1 \\ 0 \\ 1 \end{bmatrix}.$$

As the following example illustrates, the first step in finding the *vector form* for the general solution is to find the *general solution* of the system. We then write down the unknowns in vector form:

$$\mathbf{x} = \begin{bmatrix} x_1 \\ x_2 \\ \vdots \\ x_n \end{bmatrix}.$$

Next, knowing the general solution, we can enter formulas for $x_1, x_2, \ldots, x_n$ into the vector $\mathbf{x}$. Finally, we decompose the vector $\mathbf{x}$ by gathering together the independent variables. The following example illustrates the procedure.

Example 3   Let $[\,A \mid \mathbf{b}\,]$ denote the augmented matrix for a system of linear equations:

$$[A \mid \mathbf{b}] = \begin{bmatrix} 1 & -2 & 0 & 0 & 2 & 3 \\ 0 & 0 & 1 & 0 & -1 & 2 \\ 0 & 0 & 0 & 1 & 3 & -4 \end{bmatrix}.$$

Find the vector form for the general solution.

76

Solution: Since $[\,A \mid \mathbf{b}\,]$ is in reduced echelon form, we know the general solution:

$$x_1 = 3 + 2x_2 - 2x_5, \qquad x_3 = 2 + x_5, \qquad x_4 = -4 - 3x_5.$$

To express the general solution in vector form, we start with the vector of unknowns

$$\mathbf{x} = \begin{bmatrix} x_1 \\ x_2 \\ x_3 \\ x_4 \\ x_5 \end{bmatrix}.$$

Next, we use the general solution to enter formulas for the unknowns, and then decompose $\mathbf{x}$ by collecting independent variables:

$$\mathbf{x} = \begin{bmatrix} x_1 \\ x_2 \\ x_3 \\ x_4 \\ x_5 \end{bmatrix} = \begin{bmatrix} 3 + 2x_2 - 2x_5 \\ x_2 \\ 2 + x_5 \\ -4 - 3x_5 \\ x_5 \end{bmatrix} = \begin{bmatrix} 3 \\ 0 \\ 2 \\ -4 \\ 0 \end{bmatrix} + \begin{bmatrix} 2x_2 \\ x_2 \\ 0 \\ 0 \\ 0 \end{bmatrix} + \begin{bmatrix} -2x_5 \\ 0 \\ x_5 \\ -3x_5 \\ x_5 \end{bmatrix} = \begin{bmatrix} 3 \\ 0 \\ 2 \\ -4 \\ 0 \end{bmatrix} + x_2 \begin{bmatrix} 2 \\ 1 \\ 0 \\ 0 \\ 0 \end{bmatrix} + x_5 \begin{bmatrix} -2 \\ 0 \\ 1 \\ -3 \\ 1 \end{bmatrix}.$$

The last expression is the vector form for the solution. ∎

Note, in Example 3, that vector form expresses the general solution very simply as

$$\mathbf{x} = \mathbf{b} + a\mathbf{v}_1 + b\mathbf{v}_2$$

where $\mathbf{b}$, $\mathbf{v}_1$, and $\mathbf{v}_2$ are constant five-dimensional vectors and where a and b are free parameters.

Exercises 2.4

Use the following matrices and vectors in Exercises 1-8.

$$A = \begin{bmatrix} 2 & 1 \\ 1 & 3 \end{bmatrix} \quad B = \begin{bmatrix} 0 & -1 \\ 1 & 3 \end{bmatrix} \quad C = \begin{bmatrix} -2 & 3 \\ 1 & 1 \end{bmatrix}$$

$$\mathbf{r} = \begin{bmatrix} 1 \\ 0 \end{bmatrix}, \quad \mathbf{s} = \begin{bmatrix} 2 \\ -3 \end{bmatrix}, \quad \mathbf{t} = \begin{bmatrix} 1 \\ 4 \end{bmatrix}, \quad \mathbf{u} = \begin{bmatrix} -4 \\ 6 \end{bmatrix}.$$

1. Find:

   (a) $\mathbf{r} + \mathbf{s}$ (b) $2\mathbf{r} + \mathbf{t}$ (c) $2\mathbf{s} + \mathbf{u}$

2. Find:

   (a) $A\mathbf{r}$ (b) $B\mathbf{r}$ (c) $C(\mathbf{s} + 3\mathbf{t})$

In Exercises 3-6, find scalars a_1 and a_2 that satisfy the given equation, or state that the equation has no solution.

3. $a_1\mathbf{r} + a_2\mathbf{s} = \mathbf{t}$.

4. $a_1\mathbf{r} + a_2\mathbf{s} = \mathbf{u}$.

5. $a_1\mathbf{s} + a_2\mathbf{t} = \mathbf{u}$.

6. $a_1\mathbf{s} + a_2\mathbf{t} = \mathbf{r} + \mathbf{t}$.

7. Find $\mathbf{w_2}$, where $\mathbf{w_1} = B\mathbf{r}$ and $\mathbf{w_2} = A\mathbf{w_1}$. Calculate $Q = AB$. Calculate $Q\mathbf{r}$ and verify that $\mathbf{w_2}$ is equal to $Q\mathbf{r}$.

8. Find $\mathbf{w_2}$, where $\mathbf{w_1} = C\mathbf{s}$ and $\mathbf{w_2} = A\mathbf{w_1}$. Calculate $Q = AC$. Calculate $Q\mathbf{s}$ and verify that $\mathbf{w_2}$ is equal to $Q\mathbf{s}$.

9. Express each of the linear systems (i) and (ii) in the form $A\mathbf{x} = \mathbf{b}$.

(i)
$$\begin{aligned} 2x_1 - x_2 &= 3 \\ x_1 + x_2 &= 3 \end{aligned}$$

(ii)
$$\begin{aligned} x_1 - 3x_2 + x_3 &= 1 \\ x_1 - 2x_2 + x_3 &= 2 \\ x_2 - x_3 &= -1 \end{aligned}$$

10. In each of (a)-(d) determine if $\mathbf{s}$ is a solution to the matrix equation $A\mathbf{x} = \mathbf{b}$, where

$$A = \begin{bmatrix} 1 & -2 & -10 \\ -1 & 3 & 14 \\ 2 & -1 & -8 \end{bmatrix}, \quad \mathbf{x} = \begin{bmatrix} x_1 \\ x_2 \\ x_3 \end{bmatrix}, \quad \text{and } \mathbf{b} = \begin{bmatrix} 5 \\ -6 \\ 7 \end{bmatrix}.$$

(a) $\mathbf{s} = \begin{bmatrix} 1 \\ 3 \\ -1 \end{bmatrix}$ (b) $\mathbf{s} = \begin{bmatrix} 1 \\ -1 \\ 1 \end{bmatrix}$ (c) $\mathbf{s} = \begin{bmatrix} 3 \\ -1 \\ 0 \end{bmatrix}$ (d) $\mathbf{s} = \begin{bmatrix} 5 \\ -5 \\ 1 \end{bmatrix}$

11. In each of (a)-(d) determine if $\mathbf{s}$ is a solution to the matrix equation $A\mathbf{x} = \mathbf{b}$, where

$$A = \begin{bmatrix} 1 & -1 & 3 & -3 \\ 2 & -2 & 7 & -8 \\ -2 & 2 & -4 & 2 \end{bmatrix}, \quad \mathbf{x} = \begin{bmatrix} x_1 \\ x_2 \\ x_3 \\ x_4 \end{bmatrix}, \quad \text{and } \mathbf{b} = \begin{bmatrix} 10 \\ 24 \\ -12 \end{bmatrix}.$$

(a) $\mathbf{s} = \begin{bmatrix} -1 \\ 1 \\ 4 \\ 0 \end{bmatrix}$ (b) $\mathbf{s} = \begin{bmatrix} 5 \\ -1 \\ -6 \\ 1 \end{bmatrix}$ (c) $\mathbf{s} = \begin{bmatrix} -5 \\ 0 \\ 6 \\ 1 \end{bmatrix}$ (d) $\mathbf{s} = \begin{bmatrix} -4 \\ 1 \\ 6 \\ 1 \end{bmatrix}$

12. Solve the matrix equation $A\mathbf{x} = \mathbf{b}$, where

$$A = \begin{bmatrix} 1 & 2 \\ 3 & 8 \end{bmatrix}, \quad \mathbf{x} = \begin{bmatrix} x_1 \\ x_2 \end{bmatrix}, \quad \text{and } \mathbf{b} = \begin{bmatrix} 2 \\ 12 \end{bmatrix}.$$

13. Solve the matrix equation $A\mathbf{x} = \mathbf{b}$, where

$$A = \begin{bmatrix} 1 & 1 & -3 \\ -2 & -1 & 5 \\ 1 & 3 & -3 \end{bmatrix}, \quad \mathbf{x} = \begin{bmatrix} x_1 \\ x_2 \\ x_3 \end{bmatrix}, \quad \text{and } \mathbf{b} = \begin{bmatrix} -3 \\ 8 \\ 5 \end{bmatrix}.$$

14. Solve the matrix equation $A\mathbf{x} = \mathbf{b}$, where

$$A = \begin{bmatrix} 1 & 2 & 2 \\ 0 & 1 & 3 \\ 2 & 3 & 1 \end{bmatrix}, \quad \mathbf{x} = \begin{bmatrix} x_1 \\ x_2 \\ x_3 \end{bmatrix}, \quad \text{and } \mathbf{b} = \begin{bmatrix} -10 \\ -6 \\ -14 \end{bmatrix}.$$

2.5 THE MATRIX EQUATION $AX = B$

In the previous section we discussed the equation $Ax = \mathbf{b}$ where $\mathbf{x}$ and $\mathbf{b}$ are vectors. In this section we look at the general matrix equation

$$AX = B$$

where A is $(m \times n)$ and B is $(m \times q)$. A solution is an $(n \times q)$ matrix Q such that $AQ = B$. Matrix equations such as this arise quite frequently in applied problems.

The column form for a matrix

Methods for solving the matrix equation $AX = B$ are easy to describe when we use the <u>*column form*</u> to represent a matrix. In particular, let A be an $(m \times n)$ matrix. We can think of A as being made up of n columns, $\mathbf{A}_1, \mathbf{A}_2, \ldots, \mathbf{A}_n$ and we can write A as follows:

$$A = [\mathbf{A}_1, \mathbf{A}_2, \ldots, \mathbf{A}_n].$$

For example, consider the (3×4) matrix A given by

$$A = \begin{bmatrix} 2 & 0 & 7 & 1 \\ 5 & 2 & 3 & 4 \\ 1 & 2 & 5 & 1 \end{bmatrix}.$$

In column form, we have $A = [\mathbf{A}_1, \mathbf{A}_2, \mathbf{A}_3, \mathbf{A}_4]$ where

$$\mathbf{A}_1 = \begin{bmatrix} 2 \\ 5 \\ 1 \end{bmatrix}, \mathbf{A}_2 = \begin{bmatrix} 0 \\ 2 \\ 2 \end{bmatrix}, \mathbf{A}_3 = \begin{bmatrix} 7 \\ 3 \\ 5 \end{bmatrix}, \mathbf{A}_4 = \begin{bmatrix} 1 \\ 4 \\ 1 \end{bmatrix}.$$

The column form for matrix products

Having the concept of column form for a matrix, we can use it to give a simple representation for the product of two matrices; see Theorem 1.

In words, Theorem 1 says:

<u>*The kth column of AB is the product of A and the kth column of B.*</u>

Stated formally, Theorem 1 is as follows.

THEOREM 1   Let A be an $(m \times n)$ matrix and let B be an $(n \times p)$ matrix given in column form by $B = [\mathbf{B}_1, \mathbf{B}_2, \ldots, \mathbf{B}_p]$. Then, in column form, the product AB is given by

$$AB = [A\mathbf{B}_1, A\mathbf{B}_2, \ldots, A\mathbf{B}_p].$$

The proof of Theorem 1 follows directly from the definition of matrix multiplication.

Example 1   Illustrate Theorem 1 by forming AB where

$$A = \begin{bmatrix} 1 & 2 \\ 5 & -3 \end{bmatrix}, B = \begin{bmatrix} 4 & 2 & -2 \\ 5 & 4 & 1 \end{bmatrix}.$$

Solution: Using the definition of matrix multiplication, we have

$$AB = \begin{bmatrix} 1 & 2 \\ 5 & -3 \end{bmatrix}\begin{bmatrix} 4 & 2 & -2 \\ 5 & 4 & 1 \end{bmatrix} = \begin{bmatrix} 14 & 10 & 0 \\ 5 & -2 & -13 \end{bmatrix}.$$

On the other hand, the columns $A\mathbf{B}_1$, $A\mathbf{B}_2$, and $A\mathbf{B}_3$ are given by

$$A\mathbf{B}_1 = \begin{bmatrix} 1 & 2 \\ 5 & -3 \end{bmatrix}\begin{bmatrix} 4 \\ 5 \end{bmatrix} = \begin{bmatrix} 14 \\ 5 \end{bmatrix}, A\mathbf{B}_2 = \begin{bmatrix} 1 & 2 \\ 5 & -3 \end{bmatrix}\begin{bmatrix} 2 \\ 4 \end{bmatrix} = \begin{bmatrix} 10 \\ -2 \end{bmatrix}, A\mathbf{B}_3 = \begin{bmatrix} 1 & 2 \\ 5 & -3 \end{bmatrix}\begin{bmatrix} -2 \\ 1 \end{bmatrix} = \begin{bmatrix} 0 \\ -13 \end{bmatrix}.$$

Comparing AB and the matrix $[A\mathbf{B}_1, A\mathbf{B}_2, A\mathbf{B}_3]$ we see that Theorem 1 is valid for this pair of matrices A and B. ∎

Solving the matrix equation $AX = B$

We can use Theorem 1 to devise a strategy for solving matrix equations. In particular, consider the equation

$$AX = B$$

where A is $(m \times n)$ and B is $(m \times q)$.

Knowing the size of A and B, we see that X must be an $(m \times q)$ matrix. So, let us represent the unknown matrix X in column form

$$X = [\mathbf{X}_1, \mathbf{X}_2, \ldots, \mathbf{X}_q].$$

In addition, let us also represent the right-hand side matrix B in column form

$$B = [\mathbf{B}_1, \mathbf{B}_2, \ldots, \mathbf{B}_q].$$

Then, using Theorem 1, we see that the equation $AX = B$ can be rewritten as follows:

(1) $$[A\mathbf{X}_1, A\mathbf{X}_2, \ldots, A\mathbf{X}_q] = [\mathbf{B}_1, \mathbf{B}_2, \ldots, \mathbf{B}_q].$$

Thus, solving $AX = B$ is the same as solving the q different linear systems

$$A\mathbf{x} = \mathbf{B}_1$$
$$A\mathbf{x} = \mathbf{B}_2$$
$$\vdots$$
$$A\mathbf{x} = \mathbf{B}_q .$$

Example 2   Solve the matrix equation $AX = B$ where

$$A = \begin{bmatrix} 1 & 2 \\ 3 & 4 \\ -1 & -2 \end{bmatrix}, B = \begin{bmatrix} 3 & -1 \\ 5 & -1 \\ -3 & 1 \end{bmatrix}$$

and where X is (2×2).

Solution: According to (1), we can solve $AX = B$ by solving the two systems

$$A\mathbf{x} = \mathbf{B}_1 \qquad \text{and} \qquad A\mathbf{x} = \mathbf{B}_2$$

where

$$\mathbf{B}_1 = \begin{bmatrix} 3 \\ 5 \\ -3 \end{bmatrix} \qquad \text{and} \qquad \mathbf{B}_2 = \begin{bmatrix} -1 \\ -1 \\ 1 \end{bmatrix}.$$

Solving $A\mathbf{x} = \mathbf{B}_1$ by Gauss-Jordan elimination yields

$$[A \geq \mathbf{B}_1] = \begin{bmatrix} 1 & 2 & 3 \\ 3 & 4 & 5 \\ -1 & -2 & -3 \end{bmatrix} \xrightarrow[R_3+R_1]{R_2-3R_1} \begin{bmatrix} 1 & 2 & 3 \\ 0 & -2 & -4 \\ 0 & 0 & 0 \end{bmatrix} \xrightarrow[R_1-2R_2]{-(1/2)R_2} \begin{bmatrix} 1 & 0 & -1 \\ 0 & 1 & 2 \\ 0 & 0 & 0 \end{bmatrix}.$$

The solution is $x_1 = -1$ and $x_2 = 2$. Therefore, the first column of the solution matrix is

$$\mathbf{X}_1 = \begin{bmatrix} -1 \\ 2 \end{bmatrix}.$$

Next, solving $A\mathbf{x} = \mathbf{B}_2$ by Gauss-Jordan elimination yields

$$[A \ge \mathbf{B}_1] = \begin{bmatrix} 1 & 2 & -1 \\ 3 & 4 & -1 \\ -1 & -2 & 1 \end{bmatrix} \xrightarrow[R_3+R_1]{R_2-3R_1} \begin{bmatrix} 1 & 2 & -1 \\ 0 & -2 & 2 \\ 0 & 0 & 0 \end{bmatrix} \xrightarrow[R_1-2R_2]{-(1/2)R_2} \begin{bmatrix} 1 & 0 & 1 \\ 0 & 1 & -1 \\ 0 & 0 & 0 \end{bmatrix}.$$

The solution is $x_1 = 1$ and $x_2 = -1$. Therefore, the second column of the solution matrix is

$$\mathbf{X}_2 = \begin{bmatrix} 1 \\ -1 \end{bmatrix}.$$

Having found the two column vectors that make up the solution matrix $X = [\mathbf{X}_1, \mathbf{X}_2]$, we see that the equation $AX = B$ is solved by

$$X = \begin{bmatrix} -1 & 1 \\ 2 & -1 \end{bmatrix}.$$

$\blacksquare$

Solving several systems that have the same coefficient matrix

If we look at the solution to Example 2, we see that we used exactly the same row operations in reducing each of $[A \mid \mathbf{B}_1]$ and $[A \mid \mathbf{B}_2]$ to echelon form. The reason is fairly clear—the row operations required to reduce $[A \mid \mathbf{b}]$ to echelon form depend (essentially) on the coefficient matrix A and not on the right-hand side $\mathbf{b}$.

In particular, we could have worked Example 2 with less effort if we had begun with an augmented matrix $[A \mid \mathbf{B}_1, \mathbf{B}_2]$ that combined both systems. Then, after using row operations that transform A to reduced echelon form, we can read off the solutions to both systems.

The discussion above suggests a method to use when we have several systems to solve, each with the same coefficient matrix. In particular, suppose we have the following collection of linear systems to solve:

$$Ax = \mathbf{B}_1$$
$$Ax = \mathbf{B}_2$$

(2)
$$\cdot$$
$$\cdot$$
$$Ax = \mathbf{B}_k \ .$$

To solve this collection of linear systems, we can proceed as follows:

(a) Form the combined augmented matrix $C = [\, A \mid \mathbf{B}_1, \mathbf{B}_2, \ldots, \mathbf{B}_k \,]$

(b) Use row operations on C that transform the common coefficient matrix A to reduced echelon form. Then read off the solution to each system that is consistent.

Note: if the collection (2) arises from a matrix equation of the form $AX = B$ where $B = [\mathbf{B}_1, \mathbf{B}_2, \ldots, \mathbf{B}_k \,]$, then the combined augmented matrix C can be written as

$$[\, A \mid \mathbf{B}_1, \mathbf{B}_2, \ldots, \mathbf{B}_k \,] = [\, A \mid B \,].$$

We conclude this section with an example that illustrates the use of a combined augmented matrix. This example also introduces the topic of the next section, *matrix inverses*.

Example 3: Solve the matrix equation $AX = I$, where I denotes the (2×2) identity matrix and where

$$A = \begin{bmatrix} 1 & 1 \\ 2 & 3 \end{bmatrix}.$$

Solution: Solving $AX = I$ means we have to solve the two systems

$$Ax = \mathbf{I}_1 \qquad \text{and} \qquad Ax = \mathbf{I}_2$$

where

$$\mathbf{I}_1 = \begin{bmatrix} 1 \\ 0 \end{bmatrix} \qquad \text{and} \qquad \mathbf{I}_2 = \begin{bmatrix} 0 \\ 1 \end{bmatrix}.$$

Proceeding with the method suggested in **(a)** and **(b)**, we first form the combined augmented matrix $C = [\, A \mid \mathbf{I}_1, \mathbf{I}_2 \,] = [\, A \mid I \,]$ and then use row operations on C that transform A to reduced echelon form. The steps are

$$[\, A \mid I \,] = \begin{bmatrix} 1 & 1 & 1 & 0 \\ 2 & 3 & 0 & 1 \end{bmatrix} \xrightarrow{R_2 - 2R_1} \begin{bmatrix} 1 & 1 & 1 & 0 \\ 0 & 1 & -2 & 1 \end{bmatrix} \xrightarrow{R_1 - R_2} \begin{bmatrix} 1 & 0 & 3 & -1 \\ 0 & 1 & -2 & 1 \end{bmatrix}.$$

Therefore, we see the solution to $A\mathbf{x} = \mathbf{I}_1$ is $x_1 = 3$, $x_2 = -2$. Similarly, the solution to $A\mathbf{x} = \mathbf{I}_2$ is $x_1 = -1$, $x_2 = 1$.

Thus, the solution to $AX = I$ is given by

$$X = \begin{bmatrix} 3 & -1 \\ -2 & 1 \end{bmatrix}.$$

∎

Exercises 2.5

1. Let $A = \begin{bmatrix} -1 & 2 \\ 3 & 4 \\ 2 & -3 \end{bmatrix}$ and $B = \begin{bmatrix} 0 & -1 & 2 \\ 1 & 3 & 4 \end{bmatrix}$.

   (a) If $B = [\mathbf{B_1}, \mathbf{B_2}, \mathbf{B_3}]$, where $\mathbf{B_1}, \mathbf{B_2}$, and $\mathbf{B_3}$ are the columns of B, then exhibit $\mathbf{B_1}, \mathbf{B_2}$, and $\mathbf{B_3}$.

   (b) Calculate $A\mathbf{B_1}$, $A\mathbf{B_2}$, and $A\mathbf{B_3}$.

   (c) Show that the matrices $[A\mathbf{B_1}, A\mathbf{B_2}, A\mathbf{B_3}]$ and AB are equal.

2. Let A and B be the matrices defined in Exercise 1.

   (a) If $A = [\mathbf{A_1}, \mathbf{A_2}]$, where $\mathbf{A_1}$ and $\mathbf{A_2}$ are the columns of A, then exhibit $\mathbf{A_1}$ and $\mathbf{A_2}$.

   (b) Calculate $B\mathbf{A_1}$, and $B\mathbf{A_2}$.

   (c) Show that the matrices $[B\mathbf{A_1}, B\mathbf{A_2}]$ and BA are equal.

3. Let $A = \begin{bmatrix} 1 & 0 & 2 \\ 0 & 2 & -2 \\ -1 & 1 & -2 \end{bmatrix}$, $\mathbf{B_1} = \begin{bmatrix} 1 \\ 2 \\ 1 \end{bmatrix}$, $\mathbf{B_2} = \begin{bmatrix} 7 \\ -4 \\ -7 \end{bmatrix}$, $\mathbf{B_3} = \begin{bmatrix} 9 \\ -10 \\ -10 \end{bmatrix}$,

   and $\mathbf{x} = \begin{bmatrix} x_1 \\ x_2 \\ x_3 \end{bmatrix}$.

   (a) Simultaneously solve the systems of equations $A\mathbf{x} = \mathbf{B_1}$, $A\mathbf{x} = \mathbf{B_2}$, and $A\mathbf{x} = \mathbf{B_3}$ by reducing the matrix $[A|B]$, where B is the (3×3) matrix, $B = [\mathbf{B_1}, \mathbf{B_2}, \mathbf{B_3}]$.

   (b) Use the results of (a) to exhibit a (3×3) matrix C, such that $AC = B$.

87

4. Let $A = \begin{bmatrix} 1 & -1 \\ 3 & -2 \\ 1 & 1 \end{bmatrix}$, $\mathbf{B_1} = \begin{bmatrix} -2 \\ -6 \\ -2 \end{bmatrix}$, $\mathbf{B_2} = \begin{bmatrix} 2 \\ 7 \\ 4 \end{bmatrix}$, $\mathbf{B_3} = \begin{bmatrix} -4 \\ -9 \\ 2 \end{bmatrix}$, and $\mathbf{x} = \begin{bmatrix} x_1 \\ x_2 \end{bmatrix}$.

   (a) Simultaneously solve the systems of equations $A\mathbf{x} = \mathbf{B_1}$, $A\mathbf{x} = \mathbf{B_2}$, and $A\mathbf{x} = \mathbf{B_3}$ by reducing the matrix $[A|B]$, where B is the (3×3) matrix, $B = [\mathbf{B_1}, \mathbf{B_2}, \mathbf{B_3}]$.

   (b) Use the results of (a) to exhibit a (2×3) matrix C, such that $AC = B$.

5. Let $A = \begin{bmatrix} 1 & -2 \\ 3 & -4 \end{bmatrix}$ and $B = \begin{bmatrix} 3 & 2 & -1 \\ 7 & 6 & 1 \end{bmatrix}$.

   Find a (2×3) matrix C such that $AC = B$ by reducing the matrix $[A|B]$.

6. Let $A = \begin{bmatrix} 1 & 2 & 1 \\ 2 & 5 & 2 \\ 1 & 1 & 2 \end{bmatrix}$ and $B = \begin{bmatrix} 1 & 0 & 3 \\ 3 & -1 & 6 \\ 1 & 3 & 5 \end{bmatrix}$.

   Find a (3×3) matrix C such that $AC = B$ by reducing the matrix $[A|B]$.

7. Let $A = \begin{bmatrix} 1 & -1 & -5 \\ 2 & -1 & -7 \end{bmatrix}$ and $B = \begin{bmatrix} 3 & -2 \\ 5 & -1 \end{bmatrix}$.

   (a) Show that there are infinitely many (3×2) matrices C such that $AC = B$.

   (b) Exhibit a (3×2) matrix C with third row $\begin{bmatrix} 1 & 1 \end{bmatrix}$ and such that $AC = B$.

   (c) Exhibit a (3×2) matrix C with third row $\begin{bmatrix} -1 & 0 \end{bmatrix}$ and such that $AC = B$.

8. Let $A = \begin{bmatrix} 1 & 3 \\ 2 & 7 \end{bmatrix}$. Find a (2×2) matrix B such that $AB = I$ by reducing the matrix $[A|I]$.

9. Let $A = \begin{bmatrix} 2 & 3 & 7 \\ 1 & 2 & 5 \\ 3 & 7 & 19 \end{bmatrix}$. Find a (3×3) matrix B such that $AB = I$ by reducing the matrix $[A|I]$.

2.6 MATRIX INVERSES

We learned in high school algebra how to solve the equation

$$ax = b.$$

In particular, we solve $ax = b$ by multiplying both sides of the equation by a^{-1} to produce the solution

$$x = a^{-1}b.$$

In this section we look at a similar method for solving a linear system $A\mathbf{x} = \mathbf{b}$.

To illustrate the idea, let A be an $(n \times n)$ matrix. Suppose we can find an $(n \times n)$ matrix A^{-1} such that $A^{-1}A = I$, where I is the $(n \times n)$ identity matrix. If we now multiply both sides of $A\mathbf{x} = \mathbf{b}$ by A^{-1} we obtain

$$A^{-1}(A\mathbf{x}) = A^{-1}\mathbf{b},$$

or

$$(1) \qquad \qquad \mathbf{x} = A^{-1}\mathbf{b}.$$

Therefore, the existence of the matrix A^{-1} allows us to solve $A\mathbf{x} = \mathbf{b}$ in the same way that we solved $ax = b$ in high school algebra—by multiplying both sides of $A\mathbf{x} = \mathbf{b}$ by A^{-1}.

We will see an example in the next subsection.

The inverse matrix

We now give the formal definition of the *inverse matrix* for A.

DEFINITION 1   Let A be an $(n \times n)$ matrix. An $(n \times n)$ matrix, A^{-1}, is an inverse matrix for A if

$$A^{-1}A = AA^{-1} = I.$$

As we will see shortly, not every matrix has an inverse. But, if A does have an inverse, then that inverse is unique (see Theorem 1 in Section 2.8). When A has an inverse, we say that A is *invertible*. Finally, note that the concept of inverse has only been defined for square matrices.

Example 1 (a) Verify that the given matrix, A^{-1}, is an inverse matrix for A:

$$A = \begin{bmatrix} 1 & 1 \\ 2 & 3 \end{bmatrix} \qquad A^{-1} = \begin{bmatrix} 3 & -1 \\ -2 & 1 \end{bmatrix}.$$

(b) Express the following system of linear equations in matrix form as $A\mathbf{x} = \mathbf{b}$. Use A^{-1} from part (a) to solve the system (recall equation (1)):

$$\begin{aligned} x_1 + x_2 &= 5 \\ 2x_1 + 3x_2 &= 7. \end{aligned}$$

Solution: (a) To verify that A^{-1} is an inverse for A, we need to form the products $A^{-1}A$ and AA^{-1} and then check that each product is equal to I. The verification is given below:

$$A^{-1}A = \begin{bmatrix} 3 & -1 \\ -2 & 1 \end{bmatrix}\begin{bmatrix} 1 & 1 \\ 2 & 3 \end{bmatrix} = \begin{bmatrix} 1 & 0 \\ 0 & 1 \end{bmatrix}$$

$$AA^{-1} = \begin{bmatrix} 1 & 1 \\ 2 & 3 \end{bmatrix}\begin{bmatrix} 3 & -1 \\ -2 & 1 \end{bmatrix} = \begin{bmatrix} 1 & 0 \\ 0 & 1 \end{bmatrix}$$

(b) The linear system has the form $A\mathbf{x} = \mathbf{b}$ where $\mathbf{b}$ is the vector

$$\mathbf{b} = \begin{bmatrix} 5 \\ 7 \end{bmatrix}.$$

By equation (1), the solution to $A\mathbf{x} = \mathbf{b}$ is given by $\mathbf{x} = A^{-1}\mathbf{b}$. Therefore,

$$\mathbf{x} = A^{-1}\mathbf{b} = \begin{bmatrix} 3 & -1 \\ -2 & 1 \end{bmatrix}\begin{bmatrix} 5 \\ 7 \end{bmatrix} = \begin{bmatrix} 8 \\ -3 \end{bmatrix}.$$

From this we find the unique solution to the linear system is $x_1 = 8$, $x_2 = -3$. ∎

Inverses for (2 × 2) matrices

In the special case that A is a (2×2) matrix, there is a simple test for whether A is invertible. Moreover, if A is invertible, then there is a simple formula for the inverse. The details are given in Theorem 1.

THEOREM 1   Let A be a (2×2) matrix,

$$A = \begin{bmatrix} a & b \\ c & d \end{bmatrix}.$$

Let Δ be the number defined by $\Delta = ad - bc$.

(a)   If $\Delta = 0$, then A is not invertible.

(b)   If $\Delta \neq 0$, then A is invertible and A^{-1} is given by

$$A^{-1} = \frac{1}{\Delta} \begin{bmatrix} d & -b \\ -c & a \end{bmatrix}.$$

From high school algebra you probably recognize the number $\Delta = ad - bc$ as the _determinant_ of A. We will explore the connection between determinants and inverses in Chapter 4.

Example 2   Consider the system of linear equations

$$4x_1 + 2x_2 = 3$$
$$5x_1 + 5x_2 = 7 \ .$$

Express the system in the form $A\mathbf{x} = \mathbf{b}$ and use Theorem 1 to calculate A^{-1}. Next, as in equation (1), use A^{-1} to solve the system

Solution: The coefficient matrix is

$$A = \begin{bmatrix} 4 & 2 \\ 5 & 5 \end{bmatrix}.$$

Using Theorem 1 to calculate A^{-1}, we first find $\Delta = (4)(5) - (2)(5) = 10$. Thus, we have

$$A^{-1} = \frac{1}{10} \begin{bmatrix} 5 & -2 \\ -5 & 4 \end{bmatrix}.$$

Finally, using equation (1), we find the solution to the linear system is

$$\mathbf{x} = A^{-1}\mathbf{b} = \frac{1}{10} \begin{bmatrix} 5 & -2 \\ -5 & 4 \end{bmatrix} \begin{bmatrix} 3 \\ 7 \end{bmatrix} = \frac{1}{10} \begin{bmatrix} 1 \\ 13 \end{bmatrix} = \begin{bmatrix} 0.1 \\ 1.3 \end{bmatrix}.$$

Equivalently, the system has solution $x_1 = 0.1$, $x_2 = 1.3$.   ∎

Some matrices do not have inverses

Not every square matrix has an inverse. The following example illustrates this fact.

Example 3 Let A be the matrix

$$A = \begin{bmatrix} 2 & 1 \\ 4 & 2 \end{bmatrix}.$$

Show A is not invertible by showing the matrix equation $AX = I$ has no solution

Solution: The equation $AX = I$ has the form

$$\begin{bmatrix} 2 & 1 \\ 4 & 2 \end{bmatrix}\begin{bmatrix} x & u \\ y & v \end{bmatrix} = \begin{bmatrix} 1 & 0 \\ 0 & 1 \end{bmatrix}$$

or

$$\begin{bmatrix} 2x+y & 2u+v \\ 4x+2y & 4u+2v \end{bmatrix} = \begin{bmatrix} 1 & 0 \\ 0 & 1 \end{bmatrix}.$$

Comparing the first column of the matrix on the left with the first column of I, we see that x and y must satisfy the conditions

$$\begin{aligned} 2x + y &= 1 \\ 4x + 2y &= 0 \ . \end{aligned}$$

The above equations are clearly inconsistent; therefore we have shown that the matrix A has no inverse. (We could have used Theorem 1 to show that A is not invertible because the number Δ defined in Theorem 1 is equal to zero.) ∎

Finding inverses for $(n \times n)$ matrices

If A is a (2×2) matrix, then Theorem 1 gives a test for invertibility and a formula for the inverse. In this subsection we discuss an effective method for calculating A^{-1} when A is bigger than (2×2).

To find A^{-1} we need to solve the matrix equation $AX = I$. From Section 2.5, we know that an effective way to solve $AX = I$ is to form the augmented matrix

$$C = [\, A \mid I \,]$$

and then use row operations on C that transform A to reduced echelon form. In particular, we can establish the following theorem.

Note that Theorem 2 plays the same role for $(n \times n)$ matrices as Theorem 1 does for (2×2) matrices. That is, Theorem 2 gives a test for invertibility (can A be row reduced to I?) and also gives a method for calculating A^{-1} when A is invertible (perform the same row reduction steps on I).

Example 4   Let A be the (3×3) matrix

$$A = \begin{bmatrix} 1 & 3 & -1 \\ -2 & -5 & 1 \\ 1 & 5 & -2 \end{bmatrix}.$$

Use Theorem 2 to determine whether or not A is invertible. If A has an inverse, what is it?

Solution: As Theorem 2 suggests, we form the augmented matrix $[\, A \mid I \,]$ and transform it to reduced echelon form:

$$[\, A \mid I \,] = \begin{bmatrix} 1 & 3 & -1 & 1 & 0 & 0 \\ -2 & -5 & 1 & 0 & 1 & 0 \\ 1 & 5 & -2 & 0 & 0 & 1 \end{bmatrix} \xrightarrow[\;R_3 - R_1\;]{R_2 + 2R_1} \begin{bmatrix} 1 & 3 & -1 & 1 & 0 & 0 \\ 0 & 1 & -1 & 2 & 1 & 0 \\ 0 & 2 & -1 & -1 & 0 & 1 \end{bmatrix} \xrightarrow[\;R_3 - 2R_2\;]{R_1 - 3R_2}$$

$$\begin{bmatrix} 1 & 0 & 2 & -5 & -3 & 0 \\ 0 & 1 & -1 & 2 & 1 & 0 \\ 0 & 0 & 1 & -5 & -2 & 1 \end{bmatrix} \xrightarrow[\;R_2 + R_3\;]{R_1 - 2R_3} \begin{bmatrix} 1 & 0 & 0 & 5 & 1 & -2 \\ 0 & 1 & 0 & -3 & -1 & 1 \\ 0 & 0 & 1 & -5 & -2 & 1 \end{bmatrix} = [\, E \mid B \,].$$

Under the row operations above, we transformed $[\, A \mid I \,]$ to $[\, E \mid B \,]$. Since E is equal to the identity, we know by Theorem 2 that A is invertible and that B is the inverse of A. In particular,

$$A^{-1} = \begin{bmatrix} 5 & 1 & -2 \\ -3 & -1 & 1 \\ -5 & -2 & 1 \end{bmatrix}.$$

Example 5   Use the matrix A^{-1} found in Example 4 to solve the system

$$x_1 + 3x_2 - x_3 = 1$$
$$-2x_1 - 5x_2 + x_3 = -2$$
$$x_1 + 5x_2 - 2x_3 = 3$$

Solution: The system has the form $Ax = b$, where A is the matrix we inverted in Example 4. By Equation (1), the solution is $x = A^{-1}b$:

$$x = \begin{bmatrix} x_1 \\ x_2 \\ x_3 \end{bmatrix} = A^{-1}b = \begin{bmatrix} 5 & 1 & -2 \\ -3 & -1 & -1 \\ -5 & -2 & 1 \end{bmatrix}\begin{bmatrix} 1 \\ -2 \\ 3 \end{bmatrix} = \begin{bmatrix} -3 \\ 2 \\ 2 \end{bmatrix}.$$

Hence, $x_1 = -3$, $x_2 = 2$, and $x_3 = 2$ is the unique solution to this system.   ∎

Systems with a square coefficient matrix

When A is a square matrix, there is a useful connection between invertibility of A and the solutions of $Ax = b$. This connection is stated in Theorem 3, and has been illustrated in Examples 2 and 5. In words, Theorem 3 says:

When A^{-1} exists, then $Ax = b$ always has a unique solution, no matter what vector b is on the right-hand side.

(Theorem 3 actually says more than our paraphrase implies. That is, Theorem 3 also characterizes invertibility by saying: the *only* time A is invertible is when the equation $Ax = b$ always has a unique solution. But, for practical calculations, our paraphrase is the important part of Theorem 3.)

THEOREM 3   Let A be an $(n \times n)$ matrix and consider the system

$$Ax = b.$$

Then:

A is inveritble $\Leftrightarrow$ the system $Ax = b$ has a unique solution for any right-hand side b.

Exercises 2.6

In Exercises 1-4, use Theorem 1 to determine whether the given (2×2) matrix A is invertible. If A is invertible, use Theorem 1(b) to find A^{-1}. Verify that $A^{-1}A = I$ and $AA^{-1} = I$.

1. $A = \begin{bmatrix} 2 & 1 \\ 5 & 3 \end{bmatrix}$

   2. $A = \begin{bmatrix} -3 & 2 \\ 1 & 1 \end{bmatrix}$

3. $A = \begin{bmatrix} 2 & -2 \\ 2 & 3 \end{bmatrix}$

   4. $A = \begin{bmatrix} 2 & 1 \\ 4 & 2 \end{bmatrix}$

5. Show that the matrix $A = \begin{bmatrix} \sin\theta & -\cos\theta \\ \cos\theta & \sin\theta \end{bmatrix}$ has an inverse for all values of θ.

6. For what values of x is the matrix $A = \begin{bmatrix} x-1 & 4 \\ 1 & x+2 \end{bmatrix}$ not invertible.

In Exercises 7-12, reduce $[A|I]$ to find A^{-1} provided A is invertible. If A is invertible, verify that $A^{-1}A = I$.

7. $A = \begin{bmatrix} 1 & 1 \\ 2 & 3 \end{bmatrix}$

   8. $A = \begin{bmatrix} 1 & 2 \\ 2 & 1 \end{bmatrix}$

9. $A = \begin{bmatrix} 1 & 0 & 0 \\ 2 & 1 & 0 \\ 3 & 4 & 1 \end{bmatrix}$

   10. $A = \begin{bmatrix} 1 & 4 & 2 \\ 0 & 2 & 1 \\ 3 & 5 & 3 \end{bmatrix}$

11. $A = \begin{bmatrix} 1 & -2 & 3 \\ 2 & -3 & 2 \\ -1 & 0 & 5 \end{bmatrix}$

   12. $A = \begin{bmatrix} 1 & 2 & 3 & 1 \\ -1 & 0 & 2 & 1 \\ 2 & 1 & -3 & 0 \\ 1 & 1 & 2 & 1 \end{bmatrix}$

13. Use the inverse found in Exercise 7 to solve the system of equations

$$\begin{aligned} x_1 + x_2 &= 1 \\ 2x_1 + 3x_2 &= 2 \end{aligned}.$$

14. Use the inverse found in Exercise 10 to solve the system of equations

$$
\begin{aligned}
x_1 + 4x_2 + 2x_3 &= 1 \\
2x_2 + x_3 &= -1 \,. \\
3x_1 + 5x_2 + 3x_3 &= 2
\end{aligned}
$$

15. Let A be the matrix given in Exercise 7. Use the inverse found in Exercise 7 to obtain matrices B and C such that $AB = D$ and $CA = E$, where

$$
D = \begin{bmatrix} -1 & 2 & 3 \\ 1 & 0 & 2 \end{bmatrix} \quad \text{and} \quad E = \begin{bmatrix} 2 & -1 \\ 1 & 1 \\ 0 & 3 \end{bmatrix}.
$$

16. Repeat Exercise 15 where A is the matrix given in Exercise 10 and where

$$
D = \begin{bmatrix} 2 & -1 \\ 1 & 1 \\ 0 & 3 \end{bmatrix} \quad \text{and} \quad E = \begin{bmatrix} -1 & 2 & 3 \\ 1 & 0 & 2 \end{bmatrix}
$$

2.7 DETERMINANTS

In the previous section we saw that invertibility of a (2×2) matrix was determined by whether a particular number, called the *determinant*, was zero or nonzero. This section extends the determinant idea to a square matrix of any size.

The determinant of a (2×2) matrix

The determinant of a (2×2) matrix is defined as follows.

DEFINITION 1   Let A be the (2×2) matrix

$$A = \begin{bmatrix} a & b \\ c & d \end{bmatrix}.$$

The _**determinant**_ of A, denoted $\det(A)$, is defined by

$$\det(A) = ad - bc.$$

When we want to display the entries of the matrix A, we will use the following notation for $\det(A)$:

$$\det(A) = \begin{vmatrix} a & b \\ c & d \end{vmatrix} = ad - bc.$$

Example 1   Calculate the following determinants:

(a) $\begin{vmatrix} 2 & 3 \\ 1 & 4 \end{vmatrix}$ (b) $\begin{vmatrix} -4 & 2 \\ -6 & 3 \end{vmatrix}.$

Solution: We find:

(a) $\begin{vmatrix} 2 & 3 \\ 1 & 4 \end{vmatrix} = 2 \cdot 4 - 3 \cdot 1 = 5$

and

(b) $\begin{vmatrix} -4 & 2 \\ -6 & 3 \end{vmatrix} = (-4)(3) - (2)(-6) = 0.$ ■

As we will see, the determinant of an $(n \times n)$ matrix can be expressed in terms of determinants of (2×2) matrices. We begin with the idea of a *minor*. Let A denote an $(n \times n)$ matrix. By deleting row i and column j of A, we obtain a new $((n - 1) \times (n - 1))$ matrix, M_{ij}, called the ___ij-th minor___.

Example 2   Let $A = \begin{bmatrix} 2 & 1 & 7 \\ -5 & 6 & 8 \\ 0 & 4 & 3 \end{bmatrix}$. Find the minors M_{23} and M_{31}.

Solution: We form M_{23} by deleting row 2 and column 3 from A. Likewise, we form M_{31} by deleting row 3 and column 1 from A. Thus:

$$M_{23} = \begin{bmatrix} 2 & 1 \\ 0 & 4 \end{bmatrix} \quad , \quad M_{31} = \begin{bmatrix} 1 & 7 \\ 6 & 8 \end{bmatrix}.$$

∎

Cofactors

Let A denote an $(n \times n)$ matrix. The ___ij-th cofactor___ of A, denoted A_{ij}, is defined by

$$A_{ij} = (-1)^{i+j} \det(M_{ij}).$$

We will shortly define the determinant of an $(n \times n)$ matrix in terms of a "weighted sum" of cofactors.

In the definition of A_{ij}, note that the factor $(-1)^{i+j}$ is either 1 or -1, depending on whether $i + j$ is even or odd. Thus, the factor gives a sign, either + or -, to $\det(M_{ij})$.

Example 3   Let $A = \begin{bmatrix} 3 & -1 & 6 \\ 2 & 3 & 4 \\ -2 & 2 & 1 \end{bmatrix}$. Find the cofactors A_{21}, A_{22}, and A_{23}.

Solution: We obtain A_{21} as follows. First find the minor M_{21} (strike out row 2 and column 1). Then calculate $\det(M_{21})$. Finally give $\det(M_{21})$ a plus or minus sign

$$A_{21} = (-1)^{2+1} \begin{vmatrix} -1 & 6 \\ 2 & 1 \end{vmatrix} = -\begin{vmatrix} -1 & 6 \\ 2 & 1 \end{vmatrix} = -(-1-12) = 13.$$

The other cofactors are calculated in a similar fashion. In particular:

$$A_{22} = (-1)^{2+2} \begin{vmatrix} 3 & 6 \\ -2 & 1 \end{vmatrix} = (3+12) = 15 \,, \; A_{23} = (-1)^{2+3} \begin{vmatrix} 3 & -1 \\ -2 & 2 \end{vmatrix} = -(6-2) = -4.$$

∎

The determinant of an $(n \times n)$ matrix

The definition of a determinant depends on the following theorem.

THEOREM 1   Let $A = (a_{ij})$ be an $(n \times n)$ matrix. Then all the following numbers are equal:

(Cofactor expansion along row i): $a_{i1}A_{i1} + a_{i2}A_{i2} + \cdots + a_{in}A_{in}$

(Cofactor expansion along column j): $a_{1j}A_{1j} + a_{2j}A_{2j} + \cdots + a_{nj}A_{nj}$

Proof: We omit the proof of Theorem 1 since it is too time consuming. A proof can be found in most advanced linear algebra texts.   ∎

By Theorem 1 we know cofactor expansions along any row are equal, cofactor expansions along any column are equal, and any row expansion is equal to any column expansion.

The common value of these sums (a cofactor expansion by row i, or a cofactor expansion by colum j) is defined to be the *determinant* of A.

DEFINITION 2   Let $A = (a_{ij})$ be an $(n \times n)$ matrix. The **_determinant_** of A, denoted det(A), is the value found from a cofactor expansion of A.

Example 4   Find det(A) where $A = \begin{bmatrix} 3 & -1 & 6 \\ 2 & 3 & 4 \\ -2 & 2 & 1 \end{bmatrix}$ is the matrix in Example 3.

Use a second row expansion and the cofactors found in Example 3.

Solution: The cofactors found in Example 3 were the second row cofactors, $A_{21} = 13$, $A_{22} = 15$, and $A_{23} = -4$. Therefore:

$$\det(A) = a_{21}A_{21} + a_{22}A_{22} + a_{23}A_{23} = (2)(13) + (3)(15) + (4)(-4) = 55.$$

■

The next example illustrates Theorem 1 (cofactor expansions along any row or any column all have the same value).

Example 5 Let $A = \begin{bmatrix} 2 & 0 & -1 \\ -2 & 2 & 3 \\ 4 & -3 & 2 \end{bmatrix}$. Find $\det(A)$ using a row 3 expansion and then using a column 2 expansion.

Solution: The row 3 expansion gives:

$$\det(A) = (4)(1)\begin{vmatrix} 0 & -1 \\ 2 & 3 \end{vmatrix} + (-3)(-1)\begin{vmatrix} 2 & -1 \\ -2 & 3 \end{vmatrix} + (2)(1)\begin{vmatrix} 2 & 0 \\ -2 & 2 \end{vmatrix} = (8) + (12) + (8) = 28.$$

The column 2 expansion gives:

$$\det(A) = (0)(-1)\begin{vmatrix} -2 & 3 \\ 4 & 2 \end{vmatrix} + (2)(1)\begin{vmatrix} 2 & -1 \\ 4 & 2 \end{vmatrix} + (-3)(-1)\begin{vmatrix} 2 & -1 \\ -2 & 3 \end{vmatrix} = (0) + (16) + (12) = 28.$$

■

Our next example illustrates the calculation of a (4×4) determinant.

Example 6 Find $\det(A)$ where $A = \begin{bmatrix} -1 & 3 & 0 & 2 \\ 1 & -1 & 2 & 1 \\ 0 & 2 & 0 & 0 \\ -3 & 4 & -1 & 2 \end{bmatrix}$.

Solution: We can expand along any row or any column. Because of all the zeros in row 3, we choose a row 3 expansion:

$$\det(A) = 0 \cdot A_{31} + 2 \cdot A_{32} + 0 \cdot A_{33} + 0 \cdot A_{34} = (2)(-1)\begin{vmatrix} -1 & 0 & 2 \\ 1 & 2 & 1 \\ -3 & -1 & 2 \end{vmatrix}.$$

Expanding the resulting (3×3) determinant along row 1 yields:

$$\det(A) = (-2)\left((-1)(1)\begin{vmatrix} 2 & 1 \\ -1 & 2 \end{vmatrix} + (2)(1)\begin{vmatrix} 1 & 2 \\ -3 & -1 \end{vmatrix}\right) = (-2)((-5) + (10)) = -10.$$

■

Calculating determinants in practice

Finding a determinant is a computationally intensive process. In particular, using Definition 2 to compute the determinant of an $(n \times n)$ matrix will ultimately require you to find and evaluate $n!/2$ different (2×2) determinants. This is a daunting task because $n!$ is such a large number even for modest values of n. For instance, if $n = 12$, you will have to calculate 239,500,800 different (2×2) determinants in order to find a (12×12) determinant. There are determinant theorems that simplify the calculations, but it will still take you as much effort to find the determinant of an $(n \times n)$ matrix as it does to transform that matrix to reduced echelon form.

Because of the number of arithmetic operations involved, scientists and engineers normally use linear algebra software to evaluate an $(n \times n)$ determinant when n is 4 or larger.

Exercises 2.7

1. Let A be the (3×3) matrix $A = \begin{bmatrix} -2 & 1 & 4 \\ 0 & 3 & -7 \\ 1 & 2 & -5 \end{bmatrix}$.

   (a) Find all nine cofactors of A.

   (b) Use cofactor expansion along the first row to calculate det A.

   (c) Use cofactor expansion along the second row to calculate det A.

   (d) Use cofactor expansion along the first column to calculate det A.

   (e) Use the cofactors found in (a) to obtain A^{-1}.

In exercises 2-15, find the determinant of the given matrix.

2. $\begin{bmatrix} 3 & 4 \\ 1 & 3 \end{bmatrix}$
3. $\begin{bmatrix} 1 & -2 \\ 4 & -3 \end{bmatrix}$
4. $\begin{bmatrix} 2 & -4 \\ 0 & 5 \end{bmatrix}$

5. $\begin{bmatrix} 3 & -2 & 1 \\ -6 & 1 & 5 \\ 4 & 1 & 3 \end{bmatrix}$
6. $\begin{bmatrix} -2 & 5 & 0 \\ -3 & 2 & -2 \\ 1 & -3 & 0 \end{bmatrix}$
7. $\begin{bmatrix} 1 & -2 & -3 \\ 3 & 1 & 2 \\ 0 & 4 & -2 \end{bmatrix}$

8. $\begin{bmatrix} 3 & -4 & 6 \\ 0 & -2 & 1 \\ 0 & 0 & -3 \end{bmatrix}$
9. $\begin{bmatrix} 4 & 0 & 0 \\ -7 & 3 & 0 \\ 2 & 1 & -1 \end{bmatrix}$
10. $\begin{bmatrix} 5 & -7 & 4 \\ 0 & 0 & 0 \\ 2 & -4 & 6 \end{bmatrix}$

11. $\begin{bmatrix} 3 & 0 & -4 \\ -1 & 0 & 5 \\ 7 & 0 & 3 \end{bmatrix}$
12. $\begin{bmatrix} -4 & 1 & 3 & 0 \\ 2 & -2 & 1 & 4 \\ 0 & 2 & 3 & 0 \\ -3 & 2 & 0 & 0 \end{bmatrix}$
13. $\begin{bmatrix} 3 & 0 & 0 & 0 \\ 2 & 1 & 0 & 0 \\ -1 & 4 & 2 & 0 \\ 3 & 7 & 5 & -2 \end{bmatrix}$

14. $\begin{bmatrix} -5 & 6 & -1 & 4 \\ 0 & 3 & 2 & 7 \\ 0 & 0 & -1 & 4 \\ 0 & 0 & 0 & 2 \end{bmatrix}$
15. $\begin{bmatrix} 3 & -4 & 7 & 2 \\ 1 & 5 & 6 & -1 \\ 0 & 0 & 0 & 0 \\ 8 & 6 & -1 & -5 \end{bmatrix}$

In Exercises 16-18, determine all the values of λ for which the given matrix has zero determinant.

16. $\begin{bmatrix} \lambda & 8 \\ 2 & \lambda \end{bmatrix}$ 　　17. $\begin{bmatrix} \lambda+4 & -1 \\ 5 & \lambda-2 \end{bmatrix}$ 　　18. $\begin{bmatrix} \lambda-3 & 0 & 0 \\ 4 & \lambda-6 & -2 \\ 1 & 5 & \lambda+1 \end{bmatrix}$

2.8 INVERSES AND DETERMINANTS

This section states some of the algebraic properties of the inverse and explores the relationship between determinants and the existence of an inverse.

The inverse is unique

As we have seen, not every square matrix has an inverse. However, when the inverse exists, it is unique.

THEOREM 1   Let A be an invertible matrix. Then A^{-1} is unique.

Proof: Suppose B is any matrix such that $BA = AB = I$. Consider the following sequence of equalities:

$$B = BI = B(AA^{-1}) = (BA)A^{-1} = IA^{-1} = A^{-1}.$$

Thus, if $BA = AB = I$, then $B = A^{-1}$.   ∎

Inverses of products

The following theorem tells us that the product of invertible matrices is invertible and that the inverse of an invertible matrix is invertible.

THEOREM 2   Let A be an $(n \times n)$ invertible matrix. Then

(a) A^{-1} is invertible and $(A^{-1})^{-1} = A$.

(b)   If B is an $(n \times n)$ invertible matrix, then AB is also invertible and $(AB)^{-1} = B^{-1}A^{-1}$.

(c)   If k is a nonzero scalar, then kA is invertible and $(kA)^{-1} = (1/k)A^{-1}$

Proof: Part **(a)** is nothing more than a reinterpretation of the equality $AA^{-1} = A^{-1}A = I$. That is, if we multiply A^{-1} by A, we get I; therefore, the inverse of A^{-1} is the matrix A.

To prove part **(b)**, we have to verify that

$$(AB)(B^{-1}A^{-1}) = I \qquad \text{and} \qquad (B^{-1}A^{-1})(AB) = I.$$

But, as you can see, these equalities hold because matrix multiplication is associative. Therefore, the inverse of AB is $B^{-1}A^{-1}$.

To prove part **(c)**, note that $(kA)((1/k)A^{-1}) = k(1/k)AA^{-1} = I$. Similarly, we have $((1/k)A^{-1})(kA) = I$. Therefore, $(kA)^{-1} = (1/k)A^{-1}$. ∎

The following example illustrates Theorem 1.

Example 1   Let A and B be (3×3) matrices such that

$$A^{-1} = \begin{bmatrix} -1 & 0 & 3 \\ 1 & 2 & 1 \\ 0 & -3 & 2 \end{bmatrix} \quad \text{and} \quad B^{-1} = \begin{bmatrix} 2 & -1 & 0 \\ -3 & 0 & 1 \\ -1 & 1 & 0 \end{bmatrix}.$$

Find $(AB)^{-1}$ and $(2A)^{-1}$.

Solution: By Theorem 2, part **(b)**,

$$(AB)^{-1} = B^{-1}A^{-1} = \begin{bmatrix} 2 & -1 & 0 \\ -3 & 0 & 1 \\ -1 & 1 & 0 \end{bmatrix}\begin{bmatrix} -1 & 0 & 3 \\ 1 & 2 & 1 \\ 0 & -3 & 2 \end{bmatrix} = \begin{bmatrix} -3 & -2 & 5 \\ 3 & -3 & -7 \\ 2 & 2 & -2 \end{bmatrix}.$$

By Theorem 2, part **(c)**,

$$(2A)^{-1} = \frac{1}{2}A^{-1} = \frac{1}{2}\begin{bmatrix} -1 & 0 & 3 \\ 1 & 2 & 1 \\ 0 & -3 & 2 \end{bmatrix} = \begin{bmatrix} -1/2 & 0 & 3/2 \\ 1/2 & 1 & 1/2 \\ 0 & -3/2 & 1 \end{bmatrix}.$$

∎

While the product of invertible matrices is invertible, we can say nothing about the sum of invertible matrices. That is, see the exercises, the sum of two invertible matrices might be invertible or it might not be invertible. Similarly, we can say nothing about the sum of noninvertible matrices or the sum of an invertible matrix and a noninvertible matrix.

Determinants of products

Theorem 2 spoke of the inverse of a product. Theorem 3 tells us about the determinant of a product. This theorem has a number of important consequences.

THEOREM 3　　　　Let A and B be $(n \times n)$ matrices. Then

$$\det(AB) = \det(A)\det(B).$$

The proof of Theorem 3 is not hard, but it is quite long and so we omit it.

Example 2　Let A and B be $(n \times n)$ matrices such that $\det(A) = 3$ and $\det(B) = 2$. Find $\det(A^2 B)$ and $\det(AB^2)$.

Solution:　By Theorem 3,

$$\det(A^2 B) = \det(AAB) = \det(A)\det(A)\det(B) = (3)(3)(2) = 18$$

and

$$\det(AB^2) = \det(ABB) = \det(A)\det(B)\det(B) = (3)(2)(2) = 12. \quad \blacksquare$$

Example 3　Let A be a square matrix such that $A^2 = A$. What are the possible values for $\det(A)$?

Solution:　Since $A^2 = A$, we have $\det(A^2) = \det(A)$. Using Theorem 3 on the left-hand side of this equality we have

$$\det(A)\det(A) = \det(A).$$

Next, moving all terms to the left-hand side and factoring, we obtain

$$\det(A)[\det(A) - 1] = 0.$$

Thus, the only possibilities are $\det(A) = 0$ or $\det(A) = 1$. $\quad \blacksquare$

A matrix is invertible if and only if it has nonzero determinant

In Section 6 we stated an important result:

> A (2×2) matrix is invertible if and only if it has a nonzero determinant.

Theorem 4 extends this result to a square matrix of any size.

THEOREM 4   Let A be an $(n \times n)$ matrix. Then

$$A \text{ is invertible} \quad \Leftrightarrow \quad \det(A) \neq 0.$$

Moreover, if A is invertible, then $\det(A^{-1}) = 1/\det(A)$.

The proof of Theorem 4 is fairly involved and we omit it.

Example 4   Find all values x such that A is not invertible: $A = \begin{bmatrix} x+2 & 2 \\ 2 & x-1 \end{bmatrix}$.

Solution: By Theorem 4, a given matrix A is not invertible if and only if $\det(A) = 0$. So, let us calculate $\det(A)$ for the matrix in this example. We find

$$\det(A) = (x + 2)(x - 1) - 4$$

$$= x^2 + x - 6$$

$$= (x - 2)(x + 3).$$

So, A has zero determinant if and only if $x = 2$ or $x = -3$. By Theorem 4, these are the values for which A is not invertible. ∎

Exercises 2.8

In Exercises 1-6, find the inverse of the given matrix, where A and B are (3×3) matrices such that

$$A^{-1} = \begin{bmatrix} -1 & 0 & 1 \\ 2 & 1 & 0 \\ 3 & 0 & -2 \end{bmatrix} \quad \text{and} \quad B^{-1} = \begin{bmatrix} -3 & 1 & 4 \\ 1 & -1 & 0 \\ 2 & 0 & 3 \end{bmatrix}$$

1. AB
2. BA
3. A^2
4. ABA
5. $2A$
6. $\frac{1}{3}B$

In Exercises 7-10 give examples of (2×2) matrices A and B that satisfy the given condition.

7. Both A and B are invertible but $A + B$ is not.

8. Both A and B are invertible and $A + B$ is invertible.

9. Both A and B are not invertible but $A + B$ is invertible.

10. Both A and B are not invertible and $A + B$ is not invertible.

In Exercises 11-14, find the determinant of the given matrix, where A and B are $(n \times n)$ matrices such that $\det A = 2$ and $\det B = -3$.

11. A^{-1}
12. AB
13. BA
14. $A^3 B^2$

15. Let A be an $(n \times n)$ invertible matrix such that $A^2 = I$, where I is the $(n \times n)$ identity matrix. Find all possible values for $\det A$.

16. Find all values of x such that A is not invertible, where

$$A = \begin{bmatrix} -2 - x & -3 \\ 2 & 5 - x \end{bmatrix}.$$

17. For what values of x does the system of equations $A\mathbf{x} = \mathbf{b}$ have a unique solution for any right-hand side $\mathbf{b}$, where

$$A = \begin{bmatrix} x-1 & 1 & -3 \\ 0 & x & 2 \\ 0 & 1 & x \end{bmatrix}.$$

18. Let A and B be $(n \times n)$ matrices. Use Theorems 3 and 4 to show that if AB is invertible, then both A and B are invertible.

2.9 THE TRANSPOSE OF A MATRIX

The *transpose* operation, applied to a matrix A, simply interchanges the row and columns of A. This operation is one that is quite important in applications.

The definition of the transpose

We introduce the transpose with an example. Consider the matrix A given by

$$A = \begin{bmatrix} 2 & 3 & 7 \\ 1 & 4 & -6 \end{bmatrix}.$$

Let us create a new matrix, which we denote as A^T, by interchanging rows and columns of A. The new matrix is

$$A^T = \begin{bmatrix} 2 & 1 \\ 3 & 4 \\ 7 & -6 \end{bmatrix}.$$

In general, if A is an $(m \times n)$ matrix, the the *transpose* of A, denoted by A^T, is the $(n \times m)$ matrix obtained by interchanging the rows and columns of A.

Example 1 Find A^T and $\mathbf{x}^T$ where

$$A = \begin{bmatrix} 1 & 0 & 4 & 2 \\ 8 & 5 & 3 & 1 \\ 4 & 7 & 6 & 9 \end{bmatrix} \quad \text{and} \quad \mathbf{x} = \begin{bmatrix} 4 \\ 1 \\ 5 \end{bmatrix}.$$

Solution: The transposes are

$$A^T = \begin{bmatrix} 1 & 8 & 4 \\ 0 & 5 & 7 \\ 4 & 3 & 6 \\ 2 & 1 & 9 \end{bmatrix} \quad \text{and} \quad \mathbf{x}^T = \begin{bmatrix} 4 & 1 & 5 \end{bmatrix}.$$

Algebraic properties of the transpose operation

The following theorem tells how the transpose operation behaves when it is mixed in with sums, products, inverses, and determinants.

111

THEOREM 1   Let A and B be $(m \times n)$ matrices and let C be an $(m \times n)$ matrix. Then

(a) $(A + B)^T = A^T + B^T$

(b) $(AC)^T = C^T A^T$

(c) $(A^T)^T = A$

(d)   If A is a square matrix, then $\det(A^T) = \det(A)$

(e)   If A is square and invertible, then $(A^T)^{-1} = (A^{-1})^T$

Note the similarity between how the transpose operation treats products and how the inverse operation treats products. In particular, both the inverse operation and the transpose operation reverse the order of A and B:

$$(AB)^{-1} = B^{-1} A^{-1}$$

$$(AB)^T = B^T A^T$$

Symmetric matrices

There is an important class of matrices which do not change under the transpose operation. In particular, we say a square matrix A is **_symmetric_** if

$$A^T = A.$$

It is not hard to see that a symmetric matrix has to be square. Also, if $A = (a_{ij})$ is an $(n \times n)$ matrix, then A is symmetric if and only if $a_{ij} = a_{ji}$ for $1 \le i, j \le n$. Put another way, a matrix A is symmetric if the entries of A are symmetric with respect to the main diagonal. (The **_main diagonal_** of an $(n \times n)$ matrix consists of the entries $a_{11}, a_{22}, \ldots, a_{nn}$.)

Example 2   Determine which of the matrices is symmetric. Also, show that the product $C^T C$ is symmetric.

$$A = \begin{bmatrix} 2 & 5 & -3 \\ 5 & 0 & 8 \\ -3 & 8 & 4 \end{bmatrix} , \quad B = \begin{bmatrix} 1 & 3 \\ 5 & 6 \end{bmatrix} , \quad C = \begin{bmatrix} 4 & 1 & 2 \\ 7 & 0 & 8 \end{bmatrix}.$$

Solution: The matrix A is symmetric while B is not. The matrix C is not square and hence cannot be symmetric (C is (2×3) and thus C^T is (3×2)).

Even though C is not symmetric, the product $C^T C$ is symmetric. This fact can be seen from the calculation below

$$C^T C = \begin{bmatrix} 4 & 7 \\ 1 & 0 \\ 2 & 8 \end{bmatrix} \begin{bmatrix} 4 & 1 & 2 \\ 7 & 0 & 8 \end{bmatrix} = \begin{bmatrix} 65 & 4 & 64 \\ 4 & 1 & 2 \\ 64 & 2 & 68 \end{bmatrix}.$$

∎

In the exercises you are asked to show that the products QQ^T and $Q^T Q$ are always symmetric regardless of whether or not Q is symmetric.

Exercises 2.9

In each of Exercises 1-6, exhibit the transpose of the given matrix.

1. $\begin{bmatrix} -1 & 3 & 4 \\ 2 & -1 & 6 \end{bmatrix}$
     2. $\begin{bmatrix} 0 & -3 \\ -2 & 1 \\ 3 & 7 \end{bmatrix}$
     3. $\begin{bmatrix} -1 & 6 \\ 7 & 2 \end{bmatrix}$

4. $\begin{bmatrix} 0 & 3 & 2 \\ -1 & 5 & 3 \\ 7 & 2 & 5 \end{bmatrix}$
     5. $\begin{bmatrix} -1 \\ 4 \\ 3 \end{bmatrix}$
     6. $\begin{bmatrix} 1 & 0 & -2 & 1 \end{bmatrix}$

7. Verify that $(A + B)^T = A^T + B^T$ for $A = \begin{bmatrix} 2 & -3 & 2 \\ 0 & 1 & 3 \end{bmatrix}$ and $B = \begin{bmatrix} -1 & 4 & -3 \\ 2 & 1 & -1 \end{bmatrix}$.

8. Verify that $(AB)^T = B^T A^T$ for $A = \begin{bmatrix} 1 & -1 & 0 \\ 2 & 1 & 4 \end{bmatrix}$ and $B = \begin{bmatrix} -1 & 1 & 2 \\ 4 & 1 & 3 \\ -1 & 2 & -2 \end{bmatrix}$.

9. Find A^{-1} and verify that $(A^T)^{-1} = (A^{-1})^T$ for $A = \begin{bmatrix} 3 & -2 \\ 1 & -1 \end{bmatrix}$.
   [HINT: Calculate $A^T(A^{-1})^T$.]

10. Let A be a (3×3) matrix such that $A^{-1} = \begin{bmatrix} -1 & 0 & 2 \\ 4 & -1 & 1 \\ 3 & -2 & 2 \end{bmatrix}$. Exhibit $(A^T)^{-1}$.

11. For what value of x is the matrix $A = \begin{bmatrix} 0 & -2 \\ x & 3 \end{bmatrix}$ symmetric?

12. For what values of x, y, and z is the matrix $A = \begin{bmatrix} -1 & 4 & x \\ y & 0 & -3 \\ 2 & z & 6 \end{bmatrix}$ symmetric?

13. If A is an $(n \times n)$ matrix then verify that $A + A^T$ is symmetric. [HINT: Use Theorem 1, parts (a) and (c), to show that $(A+A^T)^T = A + A^T$.]

14. If A is an $(m \times n)$ matrix then verify that the product $A^T A$ is defined and show that $A^T A$ is symmetric.

15. If A is an $(m \times n)$ matrix then verify that the product AA^T is defined and show that AA^T is symmetric.

16. Show that $A^T = A^{-1}$ if $A = \begin{bmatrix} \frac{1}{\sqrt{2}} & \frac{-1}{\sqrt{2}} \\ \frac{1}{\sqrt{2}} & \frac{1}{\sqrt{2}} \end{bmatrix}$.

17. Find all possible values for $\det A$ if A is an $(n \times n)$ matrix such that $A^T = A^{-1}$.

18. Let A and B be $(n \times n)$ matrices such that B is symmetric. Show that $\det(A + B) = \det(A^T + B)$. [HINT: Consider the matrix $(A + B)^T$.]

3.1 VECTORS IN THE PLANE

The word *vector* has its origins in physics where it is used to denote a quantity having both magnitude and direction—quantities such as force, or velocity. In this chapter we will see how matrix algebra can help us solve the problems involving vectors that arise in such areas as statics and dynamics.

Three types of vectors

To minimize confusion, we will be careful to distinguish between three different types of vectors:

(a)   Physical vectors

(b)   Geometric vectors

(c)   Algebraic vectors

We will use the term *physical vector* when we refer to a physical quantity such as a force or a velocity.

In order to visualize a physical vector, we can draw a directed line segment in the *xy*-plane. We will refer to such a directed line segment as a *geometric vector*.

Finally, as we will see, any geometric vector in the plane can be represented as an ordered pair. We will call such a representation an *algebraic vector*. We will use algebraic vectors when we need to make a calculation.

Physical vectors

Some physical quantities (such as air pressure in a tire or the weight of a metal bar) can be described by a single real number. Such quantities are called *scalar quantities*.

Frequently, however, a single real number is not sufficient to describe a physical situation. For example, consider a wind blowing from the southwest toward the northeast at 10 miles per hour. This wind has both magnitude and direction and so we need two numbers to describe it.

In general, a physical quantity having both magnitude and direction is called a *vector quantity* or simply, a *vector*. Typical physical vectors are forces, displacements, velocities, and so forth.

Geometric vectors

We use geometric vectors to visualize a problem that involves physical vectors. For instance, consider Figure 1 which shows two wires pulling on a pin. If the pin tears loose from the block, which direction will it fly?

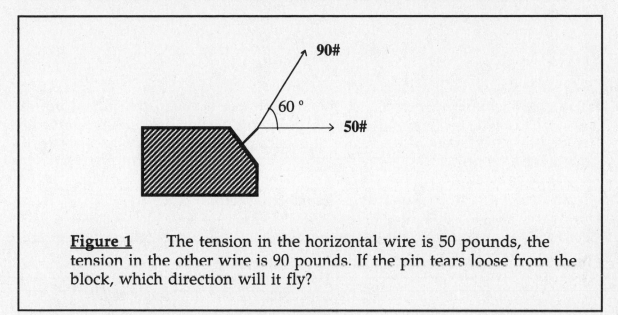

Figure 1   The tension in the horizontal wire is 50 pounds, the tension in the other wire is 90 pounds. If the pin tears loose from the block, which direction will it fly?

We can visualize the problem described in Figure 1 by sketching it in the *xy*-plane, see Figure 2. Note, in Figure 2, that the directed line segment OP representing the horizontal force has a length of 50. Similarly, OQ which represents the other force, has a length of 90.

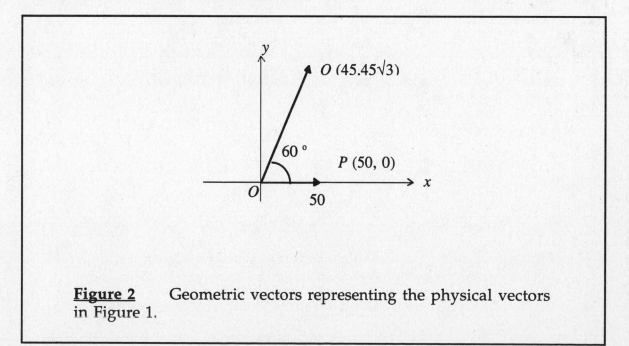

Figure 2   Geometric vectors representing the physical vectors in Figure 1.

117

The directed line segments (or arrows) in Figure 2 are examples of geometric vectors.

In general, let $A = (a_1, a_2)$ and $B = (b_1, b_2)$ be two points in the xy-plane. The directed line segment from A to B is called a ***geometric vector*** and is denoted by

$$\mathbf{v} = \overrightarrow{AB}.$$

For a given geometric vector $\overrightarrow{AB}$, the endpoint A is called the ***initial point*** while B is called the ***terminal point***. A geometric vector has both magnitude (its length) and direction—therefore, it is natural to represent physical vectors as geometric vectors.

Equality of geometric vectors

Because we are going to use geometric vectors to represent physical vectors, we adopt the following rule:

> *All geometric vectors having the same direction and magnitude will be regarded as* **equal***, regardless of whether or not they have the same endpoints.*

For example, in Figure 3, the geometric vectors $\overrightarrow{AB}$, $\overrightarrow{CD}$, and $\overrightarrow{EF}$ are all equal.

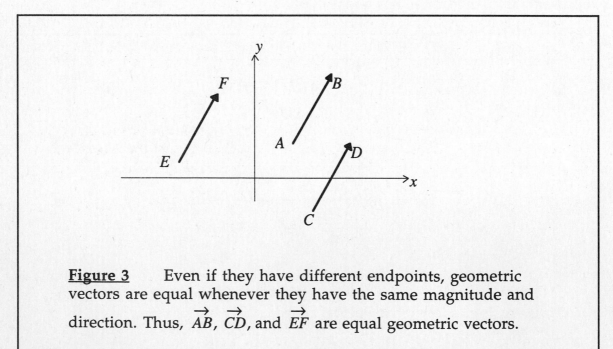

Figure 3   Even if they have different endpoints, geometric vectors are equal whenever they have the same magnitude and direction. Thus, $\overrightarrow{AB}$, $\overrightarrow{CD}$, and $\overrightarrow{EF}$ are equal geometric vectors.

118

We use this definition of equality between geometric vectors because it is agrees with the definition of equality between physical vectors.

For instance, suppose we place a brick weighing five pounds on a desk. The brick exerts the same five pound force (directed downward) no matter where we place it on the desk. Similarly, if we move the brick from the desk and place it on the floor, it still exerts a five pound force directed downward. In other words, the force vector does not change even if its point of application is changed. Geometric vectors represent physical vectors, so we do not want them to change either when they are translated in the plane.

There is a simple test for whether two geometric vectors are equal. This test is given in Theorem 1 and is based on the idea of a *position vector*.

Position vectors

Let $\mathbf{v} = \overrightarrow{AB}$ be a geometric vector with initial point $A = (a_1, a_2)$ and terminal point $B = (b_1, b_2)$, see Figure 4. Among all geometric vectors $\overrightarrow{CD}$ that are equal to $\overrightarrow{AB}$, there is exactly one of the form $\mathbf{v} = \overrightarrow{OP}$ where the initial point $O = (0, 0)$ is at the origin, see Figure 4. This vector $\overrightarrow{OP}$ is called the ***position vector*** for $\mathbf{v}$. Using geometry, it follows that the terminal point P has coordinates equal to $(b_1 - a_1, b_2 - a_2)$.

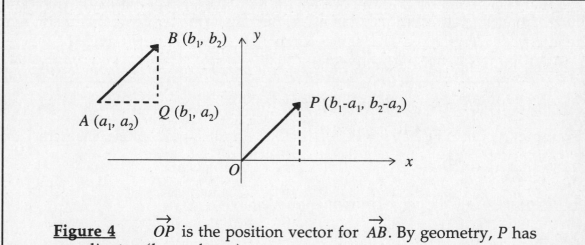

Figure 4 $\overrightarrow{OP}$ is the position vector for $\overrightarrow{AB}$. By geometry, P has coordinates $(b_1 - a_1, b_2 - a_2)$.

119

Components of a vector

Let $\mathbf{v} = \overrightarrow{OP}$ be the position vector for $\mathbf{v} = \overrightarrow{AB}$. The coordinates of P are called the *__components__* of $\mathbf{v}$. If $A = (a_1, a_2)$ and $B = (b_1, b_2)$, then we see from Figure 4:

(1a)   The x-component of $\overrightarrow{AB}$ is $b_1 - a_1$

(1b)   The y-component of $\overrightarrow{AB}$ is $b_2 - a_2$

Example 1   Let $A = (7, 3)$ and $B = (2, 5)$. Find the components of $\mathbf{v} = \overrightarrow{AB}$.

Solution: By (1), the x-component of $\overrightarrow{AB}$ is $2 - 7 = -5$ and the y-component of $\overrightarrow{AB}$ is $5 - 3 = 2$. [Note, in terms of Figure 4, the position vector $\overrightarrow{OP}$ has its terminal point at $P = (-5, 2)$.]   ■

An equality test for geometric vectors

As we see from the next theorem, two geometric vectors are equal if and only if they have the same components.

THEOREM 1   Let $\overrightarrow{AB}$ and $\overrightarrow{CD}$ be geometric vectors. Then $\overrightarrow{AB} = \overrightarrow{CD}$ if and only if their components are equal.

Proof: If two geometric vectors are equal, then they have the same position vector; hence, they have the same components. On the other hand, if two geometric vectors have the same components, then they have the same position vector; hence, they are equal.   ■

The following example illustrates Theorem 1.

Example 2   Let $A = (2,2)$, $B = (4, 5)$, $C = (3, -2)$, and $D = (5, 1)$. Use Theorem 1 to show that $\overrightarrow{AB} = \overrightarrow{CD}$.

Solution: Calculating x-components, we have

For $\overrightarrow{AB}$: $b_1 - a_1 = 4 - 2 = 2$

For $\overrightarrow{CD}$: $d_1 - c_1 = 5 - 3 = 2.$

Similarly, calculating y-components, we find

For $\overrightarrow{AB}$: $b_2 - a_2 = 5 - 2 = 3$

For $\overrightarrow{CD}$: $d_2 - c_2 = 1 - (-2) = 3$

Since both vectors have the same components, Theorem 1 tells us they are equal. ■

Moving geometric vectors about in the plane

If we know the components of a geometric vector $\mathbf{v}$, then it is easy to draw $\mathbf{v}$ using any desired initial point or terminal point.

For example, suppose we know that $\mathbf{v}$ has an x-component v_1 and has a y-component v_2. Suppose we want to represent $\mathbf{v}$ as the directed line segment $\overrightarrow{AB}$, where the initial point is $A = (a_1, a_2)$. The terminal point B is found by moving v_1 units in the x direction and then v_2 units in the y direction. In other words

(2) $$B = (a_1 + v_1, a_2 + v_2).$$

If we want to specify the terminal point B, a similar calculation is used to find the initial point A; see part **(b)** of the next example.

Example 3 A geometric vector $\mathbf{v}$ has x-component 3 and y-component -5.

(a) If we want to represent $\mathbf{v}$ in the form $\mathbf{v} = \overrightarrow{AB}$ where $A = (1, 2)$, what are the coordinates of the terminal point B?

(b) If we want to represent $\mathbf{v}$ in the form $\mathbf{v} = \overrightarrow{CD}$ where $D = (5, 7)$, what are the coordinates of the initial point C?

Solution: **(a)** By (2), the coordinates of B are $(1 + 3, 2 + (-5)) = (4, -3)$, see Figure 5.

(b) Given the coordinates of the terminal point D, we find the initial point C by moving "backwards." That is, see Figure 5,

$$C = (5 - v_1, 7 - v_2) = (5 - 3, 7 - (-5)) = (2, 12).$$

■

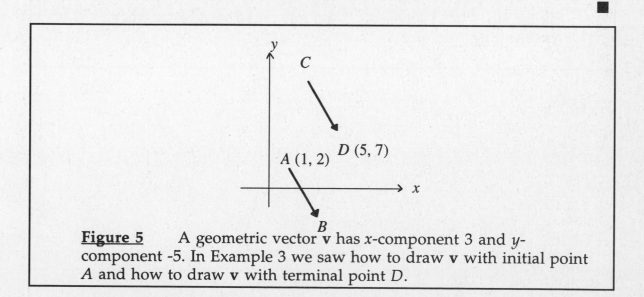

Figure 5   A geometric vector $\mathbf{v}$ has x-component 3 and y-component -5. In Example 3 we saw how to draw $\mathbf{v}$ with initial point A and how to draw $\mathbf{v}$ with terminal point D.

Our final example shows how we can use the concept of components to represent a physical vector.

Example 4 Consider a force of $\sqrt{18}$ pounds directed along a line making an angle of 45° with the horizontal, see Figure 6**(a)**. Represent this force as a geometric vector $\overrightarrow{AB}$, where the initial point A is: **(i)** $(3, 4)$, **(ii)** (a, b).

Solution: The force can be represented by any geometric vector $\overrightarrow{AB}$ that has length $\sqrt{18}$, is directed from left - to - right, and has slope equal to 1. So, see Figure 6**(b)**, the terminal point of the position vector has coordinates $P = (3, 3)$.

Knowing the components, we have:

(i) $A = (3, 4)$ and $B = (6, 7)$

(ii) $A = (a, b)$ and $B = (a + 3, b + 3)$. ■

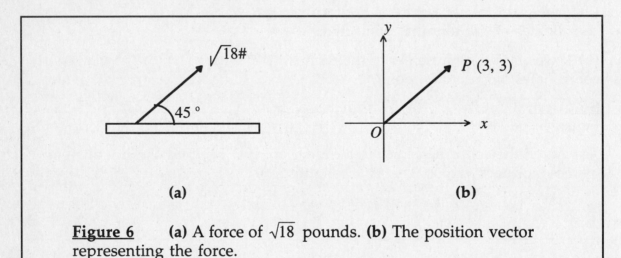

(a)   (b)

Figure 6 **(a)** A force of $\sqrt{18}$ pounds. **(b)** The position vector representing the force.

Exercises 3.1

In Exercises 1-4, graph the geometric vectors $\mathbf{u} = \overrightarrow{AB}$ and $\mathbf{v} = \overrightarrow{CD}$. Determine by inspection whether the two vectors are equal; then use Theorem 1 to verify your answer.

1. $A = (0, -2)$, $B = (-4, 3)$, $C = (1, 2)$, $D = (-3, 7)$.

2. $A = (-1, 3)$, $B = (3, -2)$, $C = (5, -1)$, $D = (1, 4)$.

3. $A = (-4, -2)$, $B = (0, 1)$, $C = (0, -2)$, $D = (3, 2)$.

4. $A = (3, 1)$, $B = (-1, -1)$, $C = (0, 3)$, $D = (-6, 0)$.

5. Let $\mathbf{u} = \overrightarrow{AB}$ and $\mathbf{v} = \overrightarrow{CD}$, where $A = (-3, 5)$, $B = (2, 2)$, $C = (3, 4)$, $D = (-2, 7)$.

    (a) Verify that $\mathbf{u}$ and $\mathbf{v}$ have the same length.

    (b) Verify that the line segments $\overline{AB}$ and $\overline{CD}$ have the same slope.

    (c) Use Theorem 1 to show that $\mathbf{u} \neq \mathbf{v}$.

    (d) Graph $\mathbf{u}$ and $\mathbf{v}$ and use the graph to explain why $\mathbf{u} \neq \mathbf{v}$.

6. Let $A = (-1, -3)$, $B = (3, 2)$, and $C = (0, 2)$. Find $D = (d_1, d_2)$ so that $\overrightarrow{AB} = \overrightarrow{CD}$.

7. Let $A = (3, -2)$, $B = (0, 3)$, and $C = (1, 0)$. Find $D = (d_1, d_2)$ so that $\overrightarrow{AB} = \overrightarrow{CD}$.

8. Let $A = (-2, 3)$, $B = (2, -1)$, and $D = (3, 2)$. Find $C = (c_1, c_2)$ so that $\overrightarrow{AB} = \overrightarrow{CD}$.

9. Let $A = (3, -2)$, $B = (3, 4)$, and $D = (-1, 7)$. Find $C = (c_1, c_2)$ so that $\overrightarrow{AB} = \overrightarrow{CD}$.

10. A force of 10 pounds is directed from left-to-right along a line at an angle of 30° (measured counter-clockwise) from the horizontal. Represent this force as a geometric vector $\overrightarrow{OP}$, where $O = (0, 0)$. Sketch a graph of $\overrightarrow{OP}$.

3.2 ALGEBRAIC VECTORS; VECTOR ADDITION

Let **v** be a geometric vector with x-component v_1 and y-component v_2. Figure 1 shows $\overrightarrow{OP}$, the position vector for **v**. Recall that the terminal point P has coordinates equal to the components of **v**; in other words,

$$P = (v_1, v_2).$$

Since **v** is completely characterized by its components, it is natural to associate **v** with the algebraic vector

$$\mathbf{v} = \begin{bmatrix} v_1 \\ v_2 \end{bmatrix}.$$

The connection between a geometric vector $\mathbf{v} = \overrightarrow{AB}$ and its algebraic representation

$$\mathbf{v} = \begin{bmatrix} v_1 \\ v_2 \end{bmatrix}$$

is so close that we will not distinguish between the two. We simply use the term _vector_ to refer to either representation. Each representation for **v** has its own advantages. The geometric representation allows us to draw diagrams as an aid for our intuition. On the other hand, the algebraic representation is very efficient when we need to make mathematical calculations.

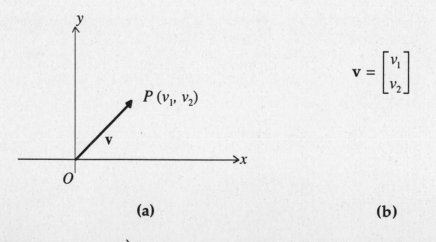

$$\mathbf{v} = \begin{bmatrix} v_1 \\ v_2 \end{bmatrix}$$

(a) **(b)**

Figure 1   **(a)** $\overrightarrow{OP}$ is the position vector for **v**; the _coordinates_ of P are equal to the components of **v**. **(b)** In the algebraic representation for **v**, the _entries_ of the column vector are the components of **v**.

A summary of the correspondence between geometric vectors and algebraic vectors

Table 1 summarizes the relationship between geometric vectors and algebraic vectors.

TABLE 1

(a) Let $\mathbf{v} = \overrightarrow{AB}$ be a geometric vector, with $A = (a_1, a_2)$ and $B = (b_1, b_2)$. Then $\mathbf{v}$ can be represented by the algebraic vector

$$\mathbf{v} = \begin{bmatrix} b_1 - a_1 \\ b_2 - a_2 \end{bmatrix}.$$

(b) Let $\mathbf{v} = \begin{bmatrix} v_1 \\ v_2 \end{bmatrix}$. Given $A = (a_1, a_2)$, $\mathbf{v}$ can be represented as the geometric vector $\overrightarrow{AB}$ where the terminal point B is given by

$$B = (a_1 + v_1, a_2 + v_2).$$

Example 1 illustrates the relationship given in Table 1.

Example 1 **(a)** Let $\mathbf{v} = \overrightarrow{AB}$ where $A = (1, 5)$ and $B = (3, 2)$. Represent $\mathbf{v}$ as an algebraic vector. **(b)** Let $\mathbf{v} = \begin{bmatrix} 4 \\ 7 \end{bmatrix}$. Represent $\mathbf{v}$ as a geometric vector with initial point $A = (5, -2)$.

Solution: **(a)** Using part **(a)** of Table 1,

$$\mathbf{v} = \begin{bmatrix} 2 \\ -3 \end{bmatrix}.$$

(b) Using part **(b)** of Table 1, $\mathbf{v} = \overrightarrow{AB}$ where $B = (5 + 4, -2 + 7) = (9, 5)$. ∎

Adding physical vectors

As you might recall from your physics or statics courses, physical vectors (forces, displacements, velocities, and so forth) can be summed.

For instance, consider Figure 2 which shows two forces, $\mathbf{F}_1$ and $\mathbf{F}_2$, applied at the same point A. We know from physics that the net effect of applying these two forces is exactly the same as applying a single force $\mathbf{F}$. This single force **F** is called the _resultant force_ and can be found by adding **F₁** and **F₂** according to the rules for adding geometric vectors.

In the next subsection we review the rules for adding geometric vectors and show how they relate to addition for algebraic vectors.

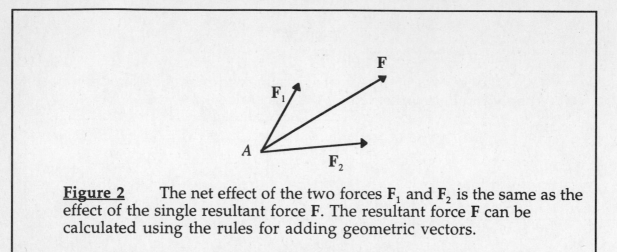

Figure 2   The net effect of the two forces F_1 and F_2 is the same as the effect of the single resultant force **F**. The resultant force **F** can be calculated using the rules for adding geometric vectors.

Adding geometric vectors

You are already familiar with the rules for adding geometric vectors from your physics courses. We review those rules now.

As in Figure 3(a), let **u** and **v** be geometric vectors. The *sum*, denoted by **u** + **v**, is the geometric vector found as follows [see Figure 3(b)]:

(a)   Translate **v** so that its initial point coincides with the terminal point of **u**.

(b)   Then, **u** + **v** is the geometric vector having the same initial point as **u** and the same terminal point as the translated vector **v**.

As will be shown shortly, it does not matter which vector is translated—we can move **v** to the tip of **u** as in Figure 3(b), or we can move **u** to the tip of **v**. We get the same sum **u** + **v** in either case.

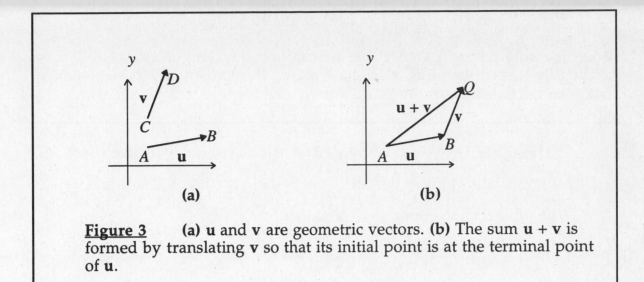

Figure 3 **(a)** **u** and **v** are geometric vectors. **(b)** The sum **u** + **v** is formed by translating **v** so that its initial point is at the terminal point of **u**.

The next example illustrates vector addition. The example also suggests that we can use algebraic vectors to calculate the sum of geometric vectors.

Example 2 Let $\mathbf{u} = \overrightarrow{AB}$ where $A = (1, 2)$ and $B = (4, 6)$, see Figure 4. Let $\mathbf{v} = \overrightarrow{CD}$ where $C = (-4, 3)$ and $D = (-2, 1)$. Find the sum **u** + **v**.

Solution: In Figure 4**(b)** we translate **v** so that its initial point is at $B = (4, 6)$. As can be seen, the terminal point of **v** is now at $Q = (6, 2)$. Thus, we see that the sum **u** + **v** is given by $\mathbf{u} + \mathbf{v} = \overrightarrow{AQ}$. ■

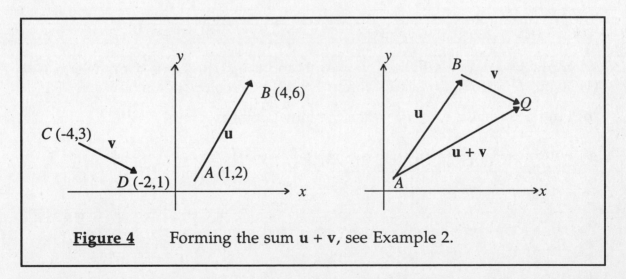

Figure 4 Forming the sum **u** + **v**, see Example 2.

If we examine the geometric construction used in Figure 4, we see that the coordinates of Q are found by adding the *components* of **v** to the *coordinates* of B. Therefore, Figure 4 suggests that we can use algebraic vectors to calculate vector sums; this idea is stated formally below, in Theorem 1.

Using algebraic vectors to calculate the sum of geometric vectors

The following theorem shows that we can find the components of $\mathbf{u} + \mathbf{v}$ by adding their corresponding algebraic vectors. In words: "the components of the sum are the sums of the components."

THEOREM 1   Let $\mathbf{u}$ and $\mathbf{v}$ be geometric vectors with algebraic representations given by $\mathbf{u} = \begin{bmatrix} u_1 \\ u_2 \end{bmatrix}$ and $\mathbf{v} = \begin{bmatrix} v_1 \\ v_2 \end{bmatrix}$. Then the sum $\mathbf{u} + \mathbf{v}$ has the following algebraic representation:

$$\mathbf{u} + \mathbf{v} = \begin{bmatrix} u_1 + v_1 \\ u_2 + v_2 \end{bmatrix}.$$

Proof: The proof is based on Figure 3 where $\mathbf{u} = \overrightarrow{AB}$. Let $A = (a_1, a_2)$ and let $B = (b_1, b_2)$. Note that, $\mathbf{u} + \mathbf{v} = \overrightarrow{AQ}$ where $Q = (b_1 + v_1, b_2 + v_2)$. Therefore, the components of $\mathbf{u} + \mathbf{v} = \overrightarrow{AQ}$ are:

The *x*-component: $(b_1 + v_1) - a_1 = (b_1 - a_1) + v_1 = u_1 + v_1$

The *y*-component: $(b_2 + v_2) - a_2 = (b_2 - a_2) + v_2 = u_2 + v_2$. ∎

The next example illustrates how we can use Theorem 1.

Example 3   Let $\mathbf{u} = \overrightarrow{AB}$ where $A = (-3, 1)$ and $B = (2, 8)$. Let $\mathbf{v} = \overrightarrow{CD}$ where $C = (1, 3)$ and $D = (4, 7)$. **(a)** Use Theorem 1 to find the components of $\mathbf{u} + \mathbf{v}$.

(b) Find S so that $\mathbf{u} + \mathbf{v} = \overrightarrow{RS}$ when the initial point is $R = (-1, 1)$.

Solution: **(a)** As algebraic vectors, $\mathbf{u} = \begin{bmatrix} 2 - (-3) \\ 8 - 1 \end{bmatrix} = \begin{bmatrix} 5 \\ 7 \end{bmatrix}$ and $\mathbf{v} = \begin{bmatrix} 4 - 1 \\ 7 - 3 \end{bmatrix} = \begin{bmatrix} 3 \\ 4 \end{bmatrix}$.

Therefore, the components of $\mathbf{u} + \mathbf{v}$ are found from $\mathbf{u} + \mathbf{v} = \begin{bmatrix} 5 \\ 7 \end{bmatrix} + \begin{bmatrix} 3 \\ 4 \end{bmatrix} = \begin{bmatrix} 8 \\ 11 \end{bmatrix}$.

(b) Having the components of $\mathbf{u} + \mathbf{v}$ we see that the terminal point S is given by $S = (-1 + 8, 1 + 11) = (7, 12)$. ∎

Exercises 3.2

In Exercises 1-4, $\mathbf{v} = \overrightarrow{AB}$ for the given points A and B. In each case, find v_1 and v_2 so that $\mathbf{v} = \begin{bmatrix} v_1 \\ v_2 \end{bmatrix}$.

1. $A = (-1, -2)$, $B = (4, 1)$.

2. $A = (1, 3)$, $B = (6, -1)$.

3. $A = (2, -1)$, $B = (-4, 4)$.

4. $A = (3, 2)$, $B = (0, -2)$.

In Exercises 5 and 6, find $B = (b_1, b_2)$ so that $\mathbf{v} = \overrightarrow{AB}$.

5. $\mathbf{v} = \begin{bmatrix} 2 \\ 5 \end{bmatrix}$, $A = (1, -2)$.

6. $\mathbf{v} = \begin{bmatrix} -3 \\ 4 \end{bmatrix}$, $A = (2, -2)$.

In Exercises 7 and 8, find $A = (a_1, a_2)$ so that $\mathbf{v} = \overrightarrow{AB}$.

7. $\mathbf{v} = \begin{bmatrix} 3 \\ -3 \end{bmatrix}$, $B = (5, 1)$.

8. $\mathbf{v} = \begin{bmatrix} -4 \\ -1 \end{bmatrix}$, $B = (-4, 2)$.

9. Let $\mathbf{u} = \begin{bmatrix} 1 \\ 3 \end{bmatrix}$, $\mathbf{v} = \begin{bmatrix} 2 \\ -2 \end{bmatrix}$, and let A denote the point $(2, -1)$.

    (a) Find points B and C so that $\mathbf{u} = \overrightarrow{AB}$, $\mathbf{v} = \overrightarrow{BC}$, and $\mathbf{u} + \mathbf{v} = \overrightarrow{AC}$.

    (b) Graph $\mathbf{u} = \overrightarrow{AB}$, $\mathbf{v} = \overrightarrow{BC}$, and $\mathbf{u} + \mathbf{v} = \overrightarrow{AC}$.

10. Repeat Exercise 9 for $\mathbf{u} = \begin{bmatrix} -2 \\ 4 \end{bmatrix}$, $\mathbf{v} = \begin{bmatrix} 3 \\ 1 \end{bmatrix}$, and $A = (0, -1)$.

11. Let $\mathbf{u} = \overrightarrow{AB}$ and $\mathbf{v} = \overrightarrow{CD}$ where $A = (-1, 2)$, $B = (3, 5)$, $C = (0, 3)$, and $D = (4, -1)$.

   (a) Find a point Q so that $\mathbf{v} = \overrightarrow{BQ}$ and $\mathbf{u} + \mathbf{v} = \overrightarrow{AQ}$.

   (b) Graph $\mathbf{u} = \overrightarrow{AB}$, $\mathbf{v} = \overrightarrow{CD}$, $\mathbf{v} = \overrightarrow{BQ}$, and $\mathbf{u} + \mathbf{v} = \overrightarrow{AQ}$.

12. Repeat Exercise 11 with $A = (-2, 4)$, $B = (1, -3)$, $C = (0, 1)$, and $D = (3, 5)$.

3.3 SCALAR MULTIPLICATION AND UNIT VECTORS

Suppose you initially push on an object with a certain force, but later decide you need to push twice as hard in the same direction. Let $\mathbf{u} = \overrightarrow{OP}$ be the position vector representing the first force, where $P = (u_1, u_2)$. Let $\mathbf{v} = \overrightarrow{OQ}$ represent the second force (see Figure 1).

Since $\mathbf{u}$ and $\mathbf{v}$ are in the same direction, Q must be on the line through O and P. But, since the second force $\mathbf{v}$ is twice as strong, the line segment OQ is twice as long as the segment OP (again, see Figure 1). Thus, the coordinates of Q are twice that of the coordinates of P; that is, $Q = (v_1, v_2) = (2u_1, 2u_2)$.

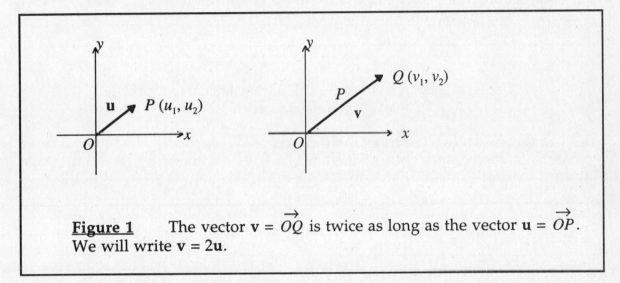

Figure 1   The vector $\mathbf{v} = \overrightarrow{OQ}$ is twice as long as the vector $\mathbf{u} = \overrightarrow{OP}$. We will write $\mathbf{v} = 2\mathbf{u}$.

Interpreting Figure 1 in terms of algebraic vectors, we have

$$\mathbf{v} = \begin{bmatrix} v_1 \\ v_2 \end{bmatrix} = \begin{bmatrix} 2u_1 \\ 2u_2 \end{bmatrix} = 2\mathbf{u}.$$

Multiplying a geometric vector by a scalar

The rules for forming the scalar multiple of a geometric vector are suggested by Figure 1. In particular, let $\mathbf{u} = \overrightarrow{AB}$ be a geometric vector and let c denote a scalar. The *scalar multiple*, denoted $c\mathbf{u}$, is defined as follows:

(a)   If $c > 0$, then $c\mathbf{u}$ is the geometric vector having the same direction as $\mathbf{u}$, but with a magnitude that is c times as large.

(b)   If $c < 0$, then $c\mathbf{u}$ is the geometric vector having the opposite direction as $\mathbf{u}$, but with a magnitude that is c times as large.

The definition of scalar multiplication is illustrated in Figure 2.

131

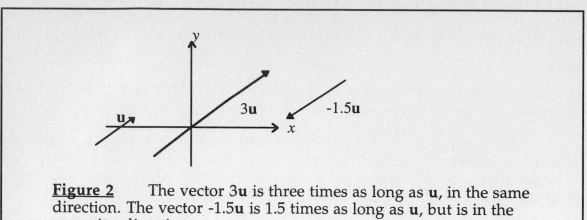

Using algebraic vectors to calculate the scalar multiple of a geometric vector

The last section stated Theorem 1 which showed how to use algebraic vectors to calculate the components of the sum, **u** + **v**. As might be expected from Figure 1, a similar result holds for scalar multiplication.

THEOREM 1   Let **u** be a geometric vector with algebraic representation $\mathbf{u} = \begin{bmatrix} u_1 \\ u_2 \end{bmatrix}$. Then the scalar multiple $c\mathbf{u}$ has the following algebraic representation:

$$c\mathbf{u} = \begin{bmatrix} cu_1 \\ cu_2 \end{bmatrix}.$$

Subtracting geometric vectors

Having defined the scalar multiple of a geometric vector, we are ready to define the difference, **u** - **v**, of two vectors. We define vector subtraction as follows:

$$\mathbf{u} - \mathbf{v} = \mathbf{u} + (-1)\mathbf{v}.$$

Vector addition and vector subtraction are often visualized in terms of the diagonals of a parallelogram, see Figure 3.

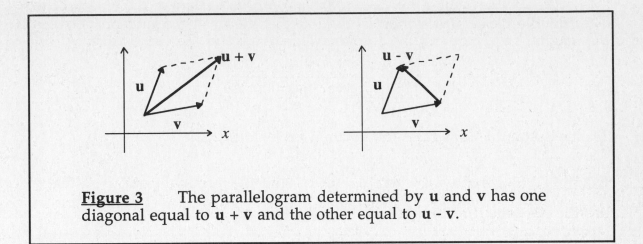

Figure 3   The parallelogram determined by **u** and **v** has one diagonal equal to **u** + **v** and the other equal to **u** - **v**.

Parallel vectors

If $c \neq 0$, then **u** and c**u** have collinear position vectors and we say they are *parallel*. In general, vectors **u** and **v** are *parallel* if there is a nonzero scalar c such that **v** = c**u**. If $c > 0$, we say **u** and **v** have the *same direction* but if $c < 0$, we say **u** and **v** have the *opposite direction*.

To determine whether or not vectors are parallel, it is easier to use the algebraic representation than the geometric representation. The next example illustrates this fact.

Example 1   Let **u** = $\overrightarrow{AB}$ and **v** = $\overrightarrow{CD}$ where $A = (1, 2)$, $B = (3, 5)$, $C = (1, 7)$, and $D = (-3, 1)$. Show that **u** and **v** are parallel vectors. Does **v** have the same direction or the opposite direction as **u**?

Solution: The algebraic representations for **u** and **v** are (respectively)

$$\mathbf{u} = \begin{bmatrix} 3-1 \\ 5-2 \end{bmatrix} = \begin{bmatrix} 2 \\ 3 \end{bmatrix} \quad \text{and} \quad \mathbf{v} = \begin{bmatrix} -3-1 \\ 1-7 \end{bmatrix} = \begin{bmatrix} -4 \\ -6 \end{bmatrix}.$$

We see that **v** = -2**u**. Hence **v** is parallel to **u**, and has the opposite direction. ■

Lengths of vectors

The length or magnitude of a geometric vector **u** = $\overrightarrow{AB}$ is the length of the line segment joining A and B. Therefore, from analytic geometry, the length of **u** = $\overrightarrow{AB}$ is equal to $\sqrt{(u_1)^2 + (u_2)^2} = \sqrt{(b_1 - a_1)^2 + (b_2 - a_2)^2}$.

We use the symbol $\|\mathbf{u}\|$, read as the *norm of u*, to denote the length of $\mathbf{u}$. The "double bar" norm symbol is used so that there is no confusion between the length of a vector and the absolute value of a scalar. To summarize:

$$\text{If } \mathbf{u} = \begin{bmatrix} u_1 \\ u_2 \end{bmatrix}, \quad \text{then} \quad \|\mathbf{u}\| = \sqrt{(u_1)^2 + (u_2)^2}\,.$$

The next example illustrates the notation for length.

Example 2   Let $\mathbf{v} = \overrightarrow{AB}$ where $A = (3, 1)$. Find the terminal point B so that $\mathbf{v}$ has the opposite direction as $\mathbf{u} = \begin{bmatrix} 1 \\ 7 \end{bmatrix}$ and where $\|\mathbf{v}\| = 10$.

Solution: Since $\mathbf{v}$ has direction opposite to $\mathbf{u}$, we know that $\mathbf{v} = c\mathbf{u}$ for some scalar $c, c < 0$. Therefore, $\mathbf{v} = \begin{bmatrix} c \\ 7c \end{bmatrix}$. Calculating $\|\mathbf{v}\|$, we find:

$$\|\mathbf{v}\| = \sqrt{(c)^2 + (7c)^2} = \sqrt{50c^2}\,.$$

So, to have $\|\mathbf{v}\| = 10$, we need $50c^2 = 100$, or $c^2 = 2$. Since $\mathbf{v}$ is in the opposite direction as $\mathbf{u}$, we need to choose the negative square root; that is, $c = -\sqrt{2}$. Thus, we obtain

$$\mathbf{v} = -\sqrt{2}\,\mathbf{u} = -\sqrt{2}\begin{bmatrix} 1 \\ 7 \end{bmatrix} = \begin{bmatrix} -\sqrt{2} \\ -7\sqrt{2} \end{bmatrix}.$$

Having the components of $\mathbf{v}$, it is now easy to find the terminal point B. In particular, $B = (3 - \sqrt{2}, 1 - 7\sqrt{2})$.   ∎

Unit vectors

If a vector $\mathbf{w}$ has length equal to 1, we say that $\mathbf{w}$ is a *unit vector*. In terms of the norm notation, $\mathbf{w}$ is a unit vector if and only if

$$\|\mathbf{w}\| = 1.$$

For example, the following vector $\mathbf{w}$ is a unit vector

$$\mathbf{w} = \begin{bmatrix} 0.6 \\ 0.8 \end{bmatrix}.$$

Finding a unit vector in the direction of a given vector

We often use a unit vector to specify _direction_. In particular, if **u** is any nonzero vector, then Equation (1) gives a formula for a unit vector **w** in the same direction as **u**:

(1)
$$\mathbf{w} = \frac{1}{\|\mathbf{u}\|}\mathbf{u}.$$

In the exercises we ask you to verify that the vector **w** defined in Equation (1) is indeed a unit vector. Then, because **w** is a positive scalar multiple of **u**, it will follow that **w** is a unit vector in the direction of **u**. (As a point of interest, - **w** is a unit vector in the direction opposite to **u**.)

If we want a vector **v** having a given length m and having the same direction as a given vector **u**, we can use Equation (1) to find it. In particular, let **w** be the unit vector defined in (1). Define **v** as follows:

$$\mathbf{v} = m\mathbf{w} = m\left(\frac{\mathbf{u}}{\|\mathbf{u}\|}\right).$$

In general, we have the following useful remark:

Remark:

Let **w** be a unit vector and let $c > 0$. Then $\mathbf{v} = c\mathbf{w}$ is a vector of length c, in the same direction as **w**.

The next example reworks a part of Example 2, using the above remark.

Example 3   Find a vector **v** having the opposite direction as $\mathbf{u} = \begin{bmatrix} 1 \\ 7 \end{bmatrix}$ and where $\|\mathbf{v}\| = 10$. (Hint: Use the above remark.)

Solution: According to the remark, we should first choose a unit vector **w** in the direction we want (opposite the direction of **u**) and then multiply **w** by 10 to produce the desired vector **v**.

Now, $\|\mathbf{u}\| = \sqrt{50}$. Therefore, using (1), a unit vector **w** having the same direction as **u** is given by

$$\mathbf{w} = \frac{1}{\|\mathbf{u}\|}\mathbf{u} = \frac{1}{\sqrt{50}}\begin{bmatrix} 1 \\ 7 \end{bmatrix}.$$

Since **w** has the same direction as **u**, we know -**w** is a unit vector in the direction opposite **u**. Multiplying -**w** by 10 will produce the vector **v** asked for in this example:

$$\mathbf{v} = -10\mathbf{w} = -\frac{10}{\sqrt{50}}\begin{bmatrix} 1 \\ 7 \end{bmatrix} = -\frac{10}{5\sqrt{2}}\begin{bmatrix} 1 \\ 7 \end{bmatrix} = -\sqrt{2}\begin{bmatrix} 1 \\ 7 \end{bmatrix}.$$

∎

The basic vectors i and j

Let $\mathbf{u} = \begin{bmatrix} u_1 \\ u_2 \end{bmatrix}$ be a vector. Then:

(2)
$$\mathbf{u} = \begin{bmatrix} u_1 \\ u_2 \end{bmatrix} = \begin{bmatrix} u_1 \\ 0 \end{bmatrix} + \begin{bmatrix} 0 \\ u_2 \end{bmatrix} = u_1\begin{bmatrix} 1 \\ 0 \end{bmatrix} + u_2\begin{bmatrix} 0 \\ 1 \end{bmatrix} = u_1\mathbf{i} + u_2\mathbf{j}$$

where the special vectors **i** and **j** are defined by

$$\mathbf{i} = \begin{bmatrix} 1 \\ 0 \end{bmatrix} \quad \text{and} \quad \mathbf{j} = \begin{bmatrix} 0 \\ 1 \end{bmatrix}.$$

As you can see, the vectors **i** and **j** are unit vectors; they are shown in Figure 4. From Equation (2) we see that **i** and **j** are _basic_ in the sense that any vector **u** can be made by adding a scalar multiple of **i** to a scalar multiple of **j**.

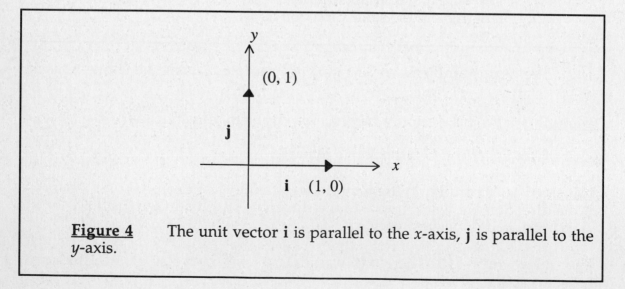

Figure 4   The unit vector **i** is parallel to the *x*-axis, **j** is parallel to the *y*-axis.

Example 4   Represent $\mathbf{u} = \begin{bmatrix} 4 \\ 9 \end{bmatrix}$ in terms of **i** and **j**.

Solution: We have

$$\mathbf{u} = \begin{bmatrix} 4 \\ 9 \end{bmatrix} = 4\begin{bmatrix} 1 \\ 0 \end{bmatrix} + 9\begin{bmatrix} 0 \\ 1 \end{bmatrix} = 4\mathbf{i} + 9\mathbf{j}.$$

■

Example 5   Let $\mathbf{u} = 2\mathbf{i} - 4\mathbf{j}$ and $\mathbf{v} = \mathbf{i} + 6\mathbf{j}$. Find the components of $\mathbf{w} = 2\mathbf{u} + \mathbf{v}$.

Solution: We find

$$\mathbf{w} = 2(2\mathbf{i} - 4\mathbf{j}) + (\mathbf{i} + 6\mathbf{j}) = 5\mathbf{i} - 2\mathbf{j} = 5\begin{bmatrix} 1 \\ 0 \end{bmatrix} - 2\begin{bmatrix} 0 \\ 1 \end{bmatrix} = \begin{bmatrix} 5 \\ -2 \end{bmatrix}$$

■

Exercises 3.3

1. Let $\mathbf{u} = \begin{bmatrix} 1 \\ 3 \end{bmatrix}$ and $\mathbf{v} = \begin{bmatrix} 2 \\ -2 \end{bmatrix}$, and let A denote the point $(-2, 1)$.

   (a) Find points B and C so that $\mathbf{u} = \overrightarrow{AB}$, $\mathbf{v} = \overrightarrow{AC}$, and $\mathbf{u} - \mathbf{v} = \overrightarrow{CB}$.

   (b) Graph $\mathbf{u} = \overrightarrow{AB}$, $\mathbf{v} = \overrightarrow{AC}$, and $\mathbf{u} - \mathbf{v} = \overrightarrow{CB}$.

2. Let $\mathbf{u} = \overrightarrow{AB}$ and $\mathbf{v} = \overrightarrow{CD}$ where $A = (-1, 2)$, $B = (3, 5)$, $C = (0, 3)$, and $D = (4, -1)$.

   (a) Find a point Q so that $\mathbf{v} = \overrightarrow{AQ}$ and $\mathbf{u} - \mathbf{v} = \overrightarrow{QB}$.

   (b) Graph $\mathbf{u} = \overrightarrow{AB}$, $\mathbf{v} = \overrightarrow{CD}$, $\mathbf{v} = \overrightarrow{AQ}$, and $\mathbf{u} - \mathbf{v} = \overrightarrow{QB}$.

3. Let $\mathbf{v} = \begin{bmatrix} 3 \\ -2 \end{bmatrix}$ and $A = (0, 5)$.

   (a) Find points B and C so that $\mathbf{v} = \overrightarrow{AB}$ and $2\mathbf{v} = \overrightarrow{AC}$.

   (b) Graph $\mathbf{v} = \overrightarrow{AB}$, $2\mathbf{v} = \overrightarrow{AC}$.

4. Let $\mathbf{v} = 2\mathbf{i} + 6\mathbf{j}$ and $A = (-2, 1)$.

   (a) Find points B and C so that $\mathbf{v} = \overrightarrow{AB}$ and $(\frac{-1}{2})\mathbf{v} = \overrightarrow{AC}$.

   (b) Graph $\mathbf{v} = \overrightarrow{AB}$ and $(\frac{-1}{2})\mathbf{v} = \overrightarrow{AC}$.

5. Let $\mathbf{v} = \overrightarrow{AB}$, where $A = (1, 4)$ and $B = (5, -1)$.

   (a) If $C = (-2, 7)$ then find D so that $2\mathbf{v} = \overrightarrow{CD}$.

   (b) Graph $\mathbf{v} = \overrightarrow{AB}$ and $2\mathbf{v} = \overrightarrow{CD}$.

6. In each of (a)-(d), find a unit vector $\mathbf{u}$ that has the same direction as the given vector $\mathbf{v}$.

(a) $\mathbf{v} = \begin{bmatrix} -1 \\ 2 \end{bmatrix}$ (b) $\mathbf{v} = \begin{bmatrix} 3 \\ 4 \end{bmatrix}$ (c) $\mathbf{v} = \mathbf{i} + \mathbf{j}$ (d) $\mathbf{v} = 3\mathbf{i} - 2\mathbf{j}$.

In Exercises 7-10, determine the terminal point B so that $\mathbf{v} = \overrightarrow{AB}$.

7. $\mathbf{v}$ has the same direction as $\begin{bmatrix} 2 \\ 1 \end{bmatrix}$, $\|\mathbf{v}\| = 4\sqrt{5}$, and $A = (-4, -2)$.

8. $\mathbf{v}$ has the opposite direction to $\begin{bmatrix} 1 \\ 3 \end{bmatrix}$, $\|\mathbf{v}\| = 3\sqrt{10}$, and $A = (4, 7)$.

9. $\mathbf{v}$ is parallel to $\mathbf{i} + 2\mathbf{j}$, $A = (3, 1)$, and B is on the y-axis.

10. $\mathbf{v}$ is parallel to $\mathbf{i} + 3\mathbf{j}$, $A = (3, 1)$, and B is on the line $y = -7$.

In Exercises 11-14, find the components of $\mathbf{u} + \mathbf{v}$ and $\mathbf{u} - 3\mathbf{v}$.

11. $\mathbf{u} = \begin{bmatrix} 1 \\ 1 \end{bmatrix}$, $\mathbf{v} = \begin{bmatrix} 1 \\ 2 \end{bmatrix}$ 12. $\mathbf{u} = \begin{bmatrix} 0 \\ 5 \end{bmatrix}$, $\mathbf{v} = \begin{bmatrix} 2 \\ -1 \end{bmatrix}$

13. $\mathbf{u} = \mathbf{i} + 2\mathbf{j}$, $\mathbf{v} = \mathbf{i} - \mathbf{j}$ 14. $\mathbf{u} = 2\mathbf{i} - \mathbf{j}$, $\mathbf{v} = 2\mathbf{i} + 2\mathbf{j}$.

15. Let $\mathbf{u} = \begin{bmatrix} a \\ b \end{bmatrix}$, where either $a \neq 0$ or $b \neq 0$, and set $\mathbf{w} = \frac{1}{\|\mathbf{u}\|}\mathbf{u}$. Verify that $\|\mathbf{w}\| = 1$.

3.4 RECTANGULAR COORDINATES IN THREE DIMENSIONS

So far, our study has been restricted to vectors in the plane; that is, to two-dimensional problems. But, since we live and work in a three-dimensional world, we need to extend vector concepts to three dimensions.

As we remember, a vector is a quantity that has both direction and magnitude. In order to discuss direction and magnitude in three dimensions, we need to have a coordinate system for three-dimensional space.

Coordinate axes in three space

There are a number of different coordinate systems we can use in three-dimensional space—for airline navigation we might use a system of geographical coordinates (latitude, longitude, and altitude).

The system we introduce here is called the ***rectangular coordinate system*** or the ***Cartesian coordinate system***. It is a direct generalization of the familiar *xy*-coordinate system for the plane.

We begin with three mutually perpendicular axes, called the *x*-axis, the *y*-axis, and the *z*-axis. We require that these axes intersect at a point designated as the ***origin***, *O*, see Figure 1.

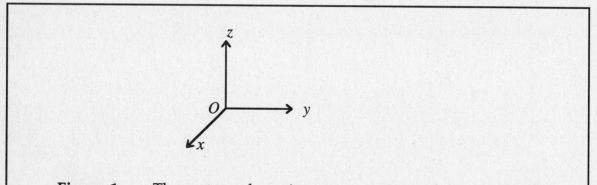

Figure 1   The rectangular axis system consists of three mutually perpendicular axes that meet at the origin, *O*.

The right-hand rule

When the coordinate axes are oriented as in Figure 1, the coordinate system is called a ***right-handed*** system. If the *x*-axis and *y*-axis are interchanged, the system is a ***left-handed*** system.

The orientation of the right-handed system can be remembered with the _**right-hand rule**_:

> _If the fingers of the right-hand are pointed in the direction of the positive x-axis and curled in the direction of the positive y-axis, then the thumb points in the direction of the positive z-axis._

We will see the right-hand rule again when we study the cross product of two vectors.

Rectangular coordinates for points in three space

Let P denote a point in space. We can locate P by moving a directed distance of a units along the x-axis, then b units along the y-axis, and then c units along the z-axis, see Figure 2. When the point P is located in this fashion, we say that P has rectangular coordinates (a, b, c).

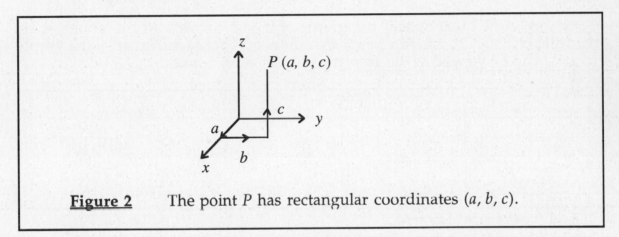

Figure 2   The point P has rectangular coordinates (a, b, c).

The plane containing the x and y axes is called the _**xy-plane**_. Similarly, the _**yz-plane**_ contains the y and z axes and the _**xz-plane**_ contains the x and z axes. Collectively, these planes are known as the _**coordinate planes**_.

The coordinate planes divide three space into eight subregions called _**octants.**_ The _**first octant**_ consists of all points $P = (a, b, c)$ where $a > 0$, $b > 0$, and $c > 0$. The other seven octants are usually not given numbers.

Example 1   Graph the points $P = (2, 3, 2)$ and $Q = (1, -2, 1)$.

Solution: The points P and Q are shown in Figure 3.   ∎

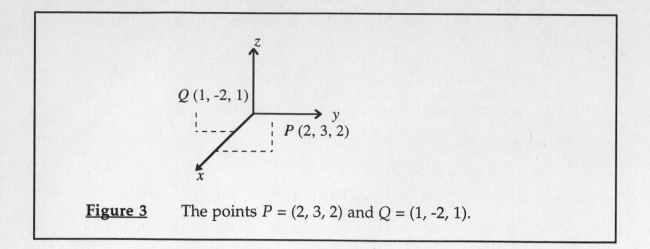

Figure 3   The points $P = (2, 3, 2)$ and $Q = (1, -2, 1)$.

Example 2   Describe the graph of the equation $z = 2$ in three space.

Solution: In three space, the graph of $z = 2$ consists of all points with coordinates $(x, y, 2)$, where x and y are arbitrary. Thus, see Figure 4, the graph of $z = 2$ can be viewed as the xy-plane translated up 2 units.

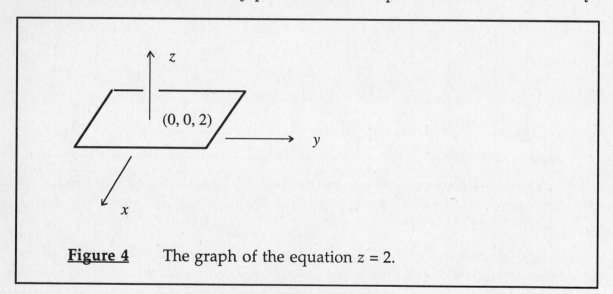

Figure 4   The graph of the equation $z = 2$.

The distance formula

Consider a geometric vector $\mathbf{u} = \overrightarrow{AB}$ in the plane. The magnitude of $\mathbf{u}$ is found by measuring the distance between the endpoints, A and B. Similarly, to define the magnitude of a vector in three space, we will need a formula for the distance between two points in space.

142

The formula for the distance between two points in space is given by:

THEOREM 1   Let $P = (p_1, p_2, p_3)$ and $Q = (q_1, q_2, q_3)$ be two points in three space. The distance between P and Q, denoted by $d(P, Q)$ is given by

$$d(P,Q) = \sqrt{(q_1 - p_1)^2 + (q_2 - p_2)^2 + (q_3 - p_3)^2}$$

Proof: The proof follows from Figure 5, after two applications of the Pythagorean formula. We ask you to supply the details in the exercises.   ■

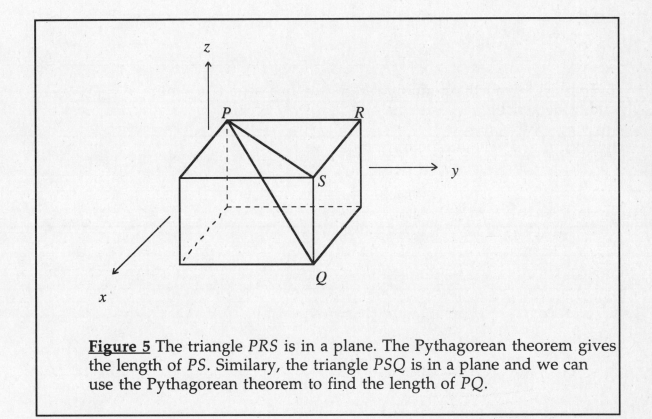

Figure 5 The triangle PRS is in a plane. The Pythagorean theorem gives the length of PS. Similary, the triangle PSQ is in a plane and we can use the Pythagorean theorem to find the length of PQ.

Example 3   Find $d(P, Q)$ where $P = (3, 2, -4)$ and $Q = (1, -1, 2)$.

Solution: By Theorem 1,

$$d(P,Q) = \sqrt{(1-3)^2 + ((-1)-2)^2 + (2-(-4))^2} = \sqrt{49} = 7$$

■

143

The midpoint formula

Having the distance formula, it is easy to develop a formula for the midpoint of a line segment. The formula is given in Theorem 2.

THEOREM 2   Let $P = (p_1, p_2, p_3)$ and $Q = (q_1, q_2, q_3)$ be two points in three space. Let M denote the midpoint of the line segment joining P and Q. Then, M is given by

$$M = \left(\frac{p_1 + q_1}{2}, \frac{p_2 + q_2}{2}, \frac{p_3 + q_3}{2} \right).$$

Proof: We leave the proof to the exercises—it requires you to establish that $d(P, M) = d(Q, M)$. ∎

Example 4   Find the midpoint, M, of the line segment joining $P = (3, 2, -4)$ and $Q = (1, -6, 2)$.

Solution: By the midpoint formula, we find

$$M = \left(\frac{3+1}{2}, \frac{2-6}{2}, \frac{-4+2}{2} \right) = (2, -2, -1)$$

∎

Exercises 3.4

In Exercises 1-4, plot the points P and Q and determine $d(P, Q)$.

1. $P = (1, 2, 1)$, $Q = (0, 2, 2)$.

2. $P = (1, 1, 0)$, $Q = (0, 0, 1)$.

3. $P = (1, 0, 0)$, $Q = (0, 0, 1)$.

4. $P = (1, 1, 1)$, $Q = (0, 0, 0)$.

In Exercises 5 and 6 find the coordinates of the midpoint of the given segment, PQ. Calculate the distance from the midpoint to the origin.

5. $P = (2, 3, 1)$, $Q = (0, 5, 7)$.

6. $P = (1, 0, 3)$, $Q = (3, 2, 5)$.

7. Let $A = (-1, 0, -3)$ and $E = (3, 6, 3)$. Find points B, C, and D on the line segment AE so that $d(A, B) = d(B, C) = d(C, D) = d(D, E) = \frac{1}{4}d(A, E)$.

In Exercises 8-12 identify the given set of points as a plane or a line in space.

8. The set of points that satisfy the equation $z = 4$.

9. The set of points with coordinates $(4, 2, z)$, z arbitrary.

10. The set of points with coordinates $(x, 2, z)$, x and z arbitrary.

11. The set of points equidistant from $(5, 0, 0)$ and $(0, 5, 0)$.

12. The set of points with coordinates (t, t, t), t arbitrary.

In Exercises 13-16, graph the given region R.

13. $R = \{(x, y, z) : |x| \leq 1, \ |y| \leq 2, \ |z| \leq 3\}$.

14. $R = \{(x, y, z) : 0 \leq x \leq 1, \ 0 \leq y \leq 2, \ z = 2\}$.

15. $R = \{(x, y, z) : 0 \leq x \leq 1, \ 0 \leq y \leq 2x, \ 0 \leq z \leq 1\}$.

145

16. $R = \{(x, y, z) : 0 \leq x, \; 0 \leq y, \; x + y \leq 1, \; 0 \leq z \leq 1\}.$

17. Let $P = (p_1, p_2, p_3)$ and $Q = (q_1, q_2, q_3)$ be two points in three space. Set $R = (p_1, q_2, p_3)$ and $S = (q_1, q_2, p_3)$. [See Figure 5.]

    (a) Apply the Pythagorean theorem to the triangle PRS to show that $d(P, S) = \sqrt{(q_1 - p_1)^2 + (q_2 - p_2)^2}$.

    (a) Apply the Pythagorean theorem to the triangle PSQ to show that $d(P, Q) = \sqrt{(q_1 - p_1)^2 + (q_2 - p_2)^2 + (q_3 - p_3)^2}$.

3.5 VECTORS IN THREE DIMENSIONS

We live and work in three-dimensional space. Hence, most physical vectors have three components. For example, consider the velocity of an airplane in flight. With respect to a rectangular coordinate system, the airplane's velocity vector will have an x component, a y component, and a z component.

The vector concepts from two dimensions can be extended directly to three dimensions. This brief section describes the extension.

Geometric vectors and their components

A physical vector can be represented by a directed line segment $\mathbf{v} = \overrightarrow{AB}$. This directed line segment is called a _geometric vector_. The length of the line segment represents the magnitude of the physical vector and the direction of the line segment represents the direction of the physical vector.

Let $\mathbf{v} = \overrightarrow{AB}$ be a geometric vector with _initial point_ $A = (a_1, a_2, a_3)$ and _terminal point_ $B = (b_1, b_2, b_3)$. The _components_ of $\mathbf{v} = \overrightarrow{AB}$ are defined for three-dimensional vectors just as they are for two-dimensional vectors. In particular:

$$\text{The } x\text{-component of } \overrightarrow{AB} \text{ is } b_1 - a_1$$
$$\text{The } y\text{-component of } \overrightarrow{AB} \text{ is } b_2 - a_2$$
$$\text{The } z\text{-component of } \overrightarrow{AB} \text{ is } b_3 - a_3$$

As before, two geometric vectors are _equal_ if they have the same components. Among all geometric vectors $\mathbf{v} = \overrightarrow{CD}$ that are equal to $\mathbf{v} = \overrightarrow{AB}$, there is exactly one of the form $\mathbf{v} = \overrightarrow{OP}$ whose initial point is at the origin. The vector $\mathbf{v} = \overrightarrow{OP}$ is the _position vector_ for $\mathbf{v}$. Because $\overrightarrow{OP}$ and $\overrightarrow{AB}$ are equal, we know that $P = (b_1 - a_1, b_2 - a_2, b_3 - a_3)$.

Algebraic vectors

Since all geometric vectors equal to $\mathbf{v} = \overrightarrow{AB}$ have the same position vector, we use the following **_algebraic vector_** to represent $\overrightarrow{AB}$ (as in two dimensions, the *components* of $\mathbf{v}$ are the *coordinates* of P):

(1)
$$\mathbf{v} = \begin{bmatrix} b_1 - a_1 \\ b_2 - a_2 \\ b_3 - a_3 \end{bmatrix}.$$

Example 1   Let $A = (1, -1, 2)$ and $B = (2, 1, 3)$. **(a)** Represent $\mathbf{v} = \overrightarrow{AB}$ as an algebraic vector. **(b)** For $C = (2, 2, 1)$, find the point D such that $\mathbf{v} = \overrightarrow{CD}$.

Solution: **(a)** Using formula (1), we have

$$\mathbf{v} = \begin{bmatrix} 2-1 \\ 1-(-1) \\ 3-2 \end{bmatrix} = \begin{bmatrix} 1 \\ 2 \\ 1 \end{bmatrix}.$$

(b) To find the terminal point D, we add the components of $\mathbf{v}$ to the coordinates of the initial point C. Thus, $D = (2 + 1, 2 + 2, 1 + 1) = (3, 4, 2)$.   ■

Addition and scalar multiplication for vectors

There is an **_addition_** and a **_scalar multiplication_** defined for geometric vectors in three space. These operations are defined exactly as they are for geometric vectors in the plane.

For instance, we find the geometric vector $\mathbf{u} + \mathbf{v}$ by translating $\mathbf{v}$ so that its initial point is at the terminal point of $\mathbf{u}$—the initial point of $\mathbf{u} + \mathbf{v}$ is the initial point of $\mathbf{u}$ while the terminal point of $\mathbf{u} + \mathbf{v}$ is the terminal point of the translated vector $\mathbf{v}$.

As with geometric vectors in the plane, these operations are easily performed using algebraic representations, see Theorem 1.

THEOREM 1   **(a)** Let $\mathbf{u} = \overrightarrow{AB}$ and $\mathbf{v} = \overrightarrow{CD}$. The algebraic representation for the geometric vector $\mathbf{u} + \mathbf{v}$ is:

$$\mathbf{u} + \mathbf{v} = \begin{bmatrix} u_1 \\ u_2 \\ u_3 \end{bmatrix} + \begin{bmatrix} v_1 \\ v_2 \\ v_3 \end{bmatrix} = \begin{bmatrix} u_1 + v_1 \\ u_2 + v_2 \\ u_3 + v_3 \end{bmatrix}.$$

(b) Let $\mathbf{u} = \overrightarrow{AB}$ and let c be a scalar. The algebraic representation for the geometric vector $c\mathbf{u}$ is:

$$c\mathbf{u} = c\begin{bmatrix} u_1 \\ u_2 \\ u_3 \end{bmatrix} = \begin{bmatrix} cu_1 \\ cu_2 \\ cu_3 \end{bmatrix}.$$

Parallel vectors, lengths of vectors, unit vectors

As before, if $\mathbf{u} = c\mathbf{v}$ for some nonzero scalar c, we say $\mathbf{u}$ is *parallel* to $\mathbf{v}$. If c is positive, then $\mathbf{u}$ has the *same direction* as $\mathbf{v}$. If c is negative, then $\mathbf{u}$ has the *opposite direction* as $\mathbf{v}$.

Let $\mathbf{u}$ be given by

$$\mathbf{u} = \begin{bmatrix} u_1 \\ u_2 \\ u_3 \end{bmatrix}.$$

The *length* of $\mathbf{u}$ (or the *norm* of $\mathbf{u}$) is defined to be:

$$\|\mathbf{u}\| = \sqrt{(u_1)^2 + (u_2)^2 + (u_3)^2}.$$

If a vector $\mathbf{w}$ has length equal to 1, we say $\mathbf{w}$ is a *unit vector*.

In applications we often need to calculate a unit vector in the direction of a given vector. This is an easy task. In particular, given a vector $\mathbf{u}$, the following vector $\mathbf{w}$ is a unit vector in the same direction as $\mathbf{u}$:

(2) $$\mathbf{w} = \frac{1}{\|\mathbf{u}\|}\mathbf{u}.$$

If we need a vector **v** having a given length, m, and in the same direction as a given vector **u**, we can find **v** with the following formula derived from equation (2).

(3)
$$\mathbf{v} = m\left(\frac{1}{\|\mathbf{u}\|}\mathbf{u}\right).$$

Example 3 Find a vector of length 7 in the direction opposite $\mathbf{u} = \begin{bmatrix} 1 \\ 2 \\ 2 \end{bmatrix}$.

Solution: A calculation shows $\|\mathbf{u}\| = 3$. Therefore, a unit vector in the direction of **u** is given by

$$\mathbf{w} = \frac{1}{3}\begin{bmatrix} 1 \\ 2 \\ 2 \end{bmatrix} = \begin{bmatrix} 1/3 \\ 2/3 \\ 2/3 \end{bmatrix}.$$

The vector -**w** is a unit vector in the direction we are interested in (-**w** is opposite **u**). Thus, the vector **v** we seek is given by **v** = -7**w**:

$$\mathbf{v} = -7\mathbf{w} = \begin{bmatrix} -7/3 \\ -14/3 \\ -14/3 \end{bmatrix}.$$

■

The basic unit vectors in three space

The basic unit vectors are denoted **i**, **j**, and **k**. In particular, **i** is along the x-axis, **j** is along the y-axis, and **k** is along the z-axis. These vectors are

$$\mathbf{i} = \begin{bmatrix} 1 \\ 0 \\ 0 \end{bmatrix}, \quad \mathbf{j} = \begin{bmatrix} 0 \\ 1 \\ 0 \end{bmatrix}, \quad \mathbf{k} = \begin{bmatrix} 0 \\ 0 \\ 1 \end{bmatrix}.$$

A given vector $\mathbf{u} = \begin{bmatrix} u_1 \\ u_2 \\ u_3 \end{bmatrix}$ can be expressed in terms of these basic unit vectors:

$$\mathbf{u} = u_1\mathbf{i} + u_2\mathbf{j} + u_3\mathbf{k}.$$

Example 3 Let $\mathbf{u} = 2\mathbf{i} - \mathbf{j} + 3\mathbf{k}$ and $\mathbf{v} = 6\mathbf{i} + 4\mathbf{j} + \mathbf{k}$. Find a unit vector in the direction of $\mathbf{u} - \mathbf{v}$.

Solution: The vector $\mathbf{u} - \mathbf{v}$ is given by $\mathbf{u} - \mathbf{v} = -4\mathbf{i} - 5\mathbf{j} + 2\mathbf{k} = \begin{bmatrix} -4 \\ -5 \\ 2 \end{bmatrix}$. The length

of $\mathbf{u} - \mathbf{v}$ is $\sqrt{45} = 3\sqrt{5}$. Thus a unit vector in the direction of $\mathbf{u} - \mathbf{v}$ is

$$\mathbf{w} = \frac{1}{\|\mathbf{u} - \mathbf{v}\|}(\mathbf{u} - \mathbf{v}) = \frac{1}{3\sqrt{5}}\begin{bmatrix} -4 \\ -5 \\ 2 \end{bmatrix}.$$

$\blacksquare$

Exercises 3.5

In Exercises 1-4, the vector $\mathbf{v} = \overrightarrow{AB}$ has given endpoints A and B.

   (a)   Give the algebraic representation for $\mathbf{v}$.

   (b)   If $C = (-1, 2, 1)$, then find D so that $\mathbf{v} = \overrightarrow{CD}$.

1. $A = (2, 0, 7)$, $B = (0, 3, 4)$.

2. $A = (-1, 2, 3)$, $B = (2, 4, 0)$.

3. $A = (2, 4, 0)$, $B = (-1, 2, 3)$.

4. $A = (1, -2, 4)$, $B = (1, 3, -3)$.

In Exercises 5-8 determine a point A such that $\mathbf{v} = \overrightarrow{AB}$, where $B = (4, 3, 2)$.

5. $\mathbf{v} = \begin{bmatrix} 0 \\ 3 \\ 2 \end{bmatrix}$ 6. $\mathbf{v} = \begin{bmatrix} 1 \\ 1 \\ 1 \end{bmatrix}$ 7. $\mathbf{v} = 2\mathbf{j}$ 8. $\mathbf{v} = 6\mathbf{i} + 3\mathbf{k}$

In Exercises 9-12, vectors $\mathbf{u}$ and $\mathbf{v}$ are given. Find: (a) $\mathbf{u} + 2\mathbf{v}$, (b) $\|\mathbf{u} - \mathbf{v}\|$, (c) a vector $\mathbf{w}$ such that $\mathbf{u} + 2\mathbf{w} = \mathbf{v}$.

9. $\mathbf{u} = \begin{bmatrix} 1 \\ 3 \\ 2 \end{bmatrix}, \mathbf{v} = \begin{bmatrix} 4 \\ 3 \\ 6 \end{bmatrix}$ 10. $\mathbf{u} = \begin{bmatrix} 1 \\ 3 \\ 2 \end{bmatrix}, \mathbf{v} = \begin{bmatrix} 4 \\ 3 \\ 6 \end{bmatrix}$

11. $\mathbf{u} = 9\mathbf{i} - 3\mathbf{j} + 2\mathbf{k}$, $\mathbf{v} = \mathbf{i} + \mathbf{k}$ 12. $\mathbf{u} = -5\mathbf{i} + 7\mathbf{j}$, $\mathbf{v} = 2\mathbf{i} - 3\mathbf{j} + \mathbf{k}$

In Exercises 13-18, determine a vector $\mathbf{u}$ that satisfies the given conditions.

13. $\mathbf{u}$ has the same direction as $\mathbf{v} = \mathbf{i} + \mathbf{j}$ and $\|\mathbf{u}\| = \sqrt{8}$.

14. $\mathbf{u}$ has the same direction as $\mathbf{k}$ and $\|\mathbf{u}\| = \sqrt{4}$.

15. **u** has opposite direction to $\mathbf{v} = \begin{bmatrix} 1 \\ 0 \\ 1 \end{bmatrix}$ and $\|\mathbf{u}\| = \sqrt{32}$.

16. **u** has opposite direction to $\mathbf{v} = \begin{bmatrix} -1 \\ 2 \\ 2 \end{bmatrix}$ and $\|\mathbf{u}\| = 5$.

17. **u** is parallel to $\mathbf{v} = \begin{bmatrix} 1 \\ 2 \\ 0 \end{bmatrix}$ and $\mathbf{v} = \overrightarrow{AB}$, where $A = (1,1,1)$ and B is in the xz-plane.

18. **u** is parallel to $\mathbf{v} = \begin{bmatrix} 4 \\ 8 \\ 2 \end{bmatrix}$ and $\mathbf{v} = \overrightarrow{AB}$, where $A = (1,1,1)$ and B is in the xy-plane.

3.6 THE DOT PRODUCT

Applications (such as calculating the work done by a force) often require us to determine the angle between two geometric vectors. Figure 1 shows the angle θ between two vectors **u** and **v**. As we will see, θ can be found using the *dot product*, the topic of this section.

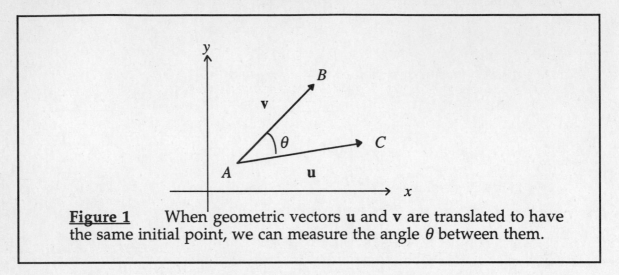

Figure 1   When geometric vectors **u** and **v** are translated to have the same initial point, we can measure the angle θ between them.

The dot product of two vectors

To find the angle θ in Figure 1, we consider the triangle ABC and interpret the vectors $\mathbf{v} = \overrightarrow{AB}$, $\mathbf{u} = \overrightarrow{AC}$, and $\mathbf{u} - \mathbf{v} = \overrightarrow{BC}$ as the sides of ABC (see Figure 2).

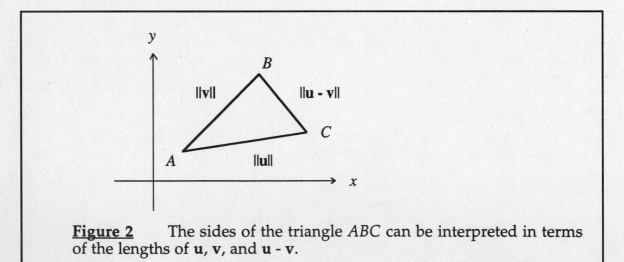

Figure 2   The sides of the triangle ABC can be interpreted in terms of the lengths of **u**, **v**, and **u** - **v**.

Combining Figure 2 and the *Law of Cosines*, we obtain an expression involving $\cos\theta$:

(1) $$\|\mathbf{u}-\mathbf{v}\|^2 = \|\mathbf{u}\|^2 + \|\mathbf{v}\|^2 - 2\|\mathbf{u}\|\,\|\mathbf{v}\|\cos\theta$$

To solve (1) for cos θ, let **u** and **v** be given by

$$\mathbf{u} = \begin{bmatrix} u_1 \\ u_2 \end{bmatrix}, \quad \mathbf{v} = \begin{bmatrix} v_1 \\ v_2 \end{bmatrix}.$$

We ask you to insert **u** and **v** into Equation (1) and simplify, finding:

(2a) $$u_1v_1 + u_2v_2 = \|\mathbf{u}\|\|\mathbf{v}\|\cos\theta.$$

Then, solving Equation (2a) for cos θ, we arrive at

$$\cos\theta = \frac{u_1v_1 + u_2v_2}{\|\mathbf{u}\|\|\mathbf{v}\|}.$$

If we now take the inverse cosine of the above expression, we will have the angle θ. The numerator of the above expression is called the *dot product* of **u** and **v**. (For θ to be defined, both **u** and **v** must be *nonzero* vectors.)

The definition of the dot product

Although it is a bit more complicated, the calculations leading from Equation (1) to Equation (2a) can be carried out in three dimensions as well. In fact, Equation (2a) becomes (in three dimensions)

(2b) $$u_1v_1 + u_2v_2 + u_3v_3 = \|\mathbf{u}\|\,\|\mathbf{v}\|\cos\theta.$$

The left-hand side of (2b) defines the dot product in three space. In general,

---

DEFINITION 1   **(a)** If **u** and **v** are two-dimensional vectors, then the *dot product* of **u** and **v**, denoted **u**•**v**, is defined by

$$\mathbf{u}\bullet\mathbf{v} = u_1v_1 + u_2v_2.$$

(b) If **u** and **v** are three-dimensional vectors, then

$$\mathbf{u}\bullet\mathbf{v} = u_1v_1 + u_2v_2 + u_3v_3.$$

The angle between two vectors

We use the dot product notation to summarize Equations (2a) and (2b).

THEOREM 1   Let **u** and **v** be two nonzero vectors in the plane or in three space. If θ denotes the angle between **u** and **v**, then

$$\mathbf{u} \bullet \mathbf{v} = \|\mathbf{u}\| \|\mathbf{v}\| \cos\theta \, .$$

The next example illustrates the use of Theorem 1.

Example 1   Let $\mathbf{u} = \begin{bmatrix} 1 \\ 3 \\ -2 \end{bmatrix}$ and $\mathbf{v} = \begin{bmatrix} 4 \\ -1 \\ -3 \end{bmatrix}$. **(a)** Calculate **u**•**v**. **(b)** Using Theorem 1, find the angle between **u** and **v**.

Solution: **(a)** The number **u**•**v** is given by

$$\mathbf{u} \bullet \mathbf{v} = (1)(4) + (3)(-1) + (-2)(-3) = 4 - 3 + 6 = 7.$$

(b) Solving for $\cos\theta$ in Theorem 1, we have

$$\cos\theta = \frac{\mathbf{u} \bullet \mathbf{v}}{\|\mathbf{u}\| \|\mathbf{v}\|} = \frac{7}{\sqrt{14}\sqrt{26}} = \frac{7}{\sqrt{364}} = 0.366899....$$

Taking the inverse cosine, we find $\theta = 1.1951...$ radians (68.4755... degrees). ∎

Algebraic properties of the dot product

The following theorem lists selected algebraic properties of the dot product.

THEOREM 2   Let **u**, **v**, and **w** denote vectors in space or in the plane, and let c denote a scalar. Then:

(a) $\mathbf{u} \bullet \mathbf{u} \geq 0$ **(b)** $\mathbf{u} \bullet \mathbf{v} = \mathbf{v} \bullet \mathbf{u}$

(c) $\mathbf{u} \bullet (c\mathbf{v}) = c(\mathbf{u} \bullet \mathbf{v})$ **(d)** $\mathbf{u} \bullet (\mathbf{v} + \mathbf{w}) = \mathbf{u} \bullet \mathbf{v} + \mathbf{u} \bullet \mathbf{w}$

We leave the verification of Theorem 2 to the exercises.

Orthogonal vectors

Let **u** and **v** be nonzero vectors and let θ denote the angle between them. When $\theta = \pi/2$, we say that **u** and **v** are _perpendicular_ or _orthogonal._ By Theorem 1, we see that **u** and **v** are orthogonal if and only if

$$\mathbf{u} \bullet \mathbf{v} = 0.$$

In the plane, the basic unit vectors **i** and **j** are orthogonal. In three space, the basic unit vectors **i**, **j**, and **k** are mutually orthogonal. The next example also illustrates the concept of othogonality.

Example 2   Let $\mathbf{u} = \begin{bmatrix} 2 \\ -3 \\ 7 \end{bmatrix}$, $\mathbf{v} = \begin{bmatrix} 1 \\ 3 \\ 1 \end{bmatrix}$ and $\mathbf{w} = \begin{bmatrix} 3 \\ 2 \\ 0 \end{bmatrix}$. Show that **u** is orthogonal to both **v** and **w**, but that **v** and **w** are not orthogonal.

Solution: Forming dot products, we find

$$\mathbf{u} \bullet \mathbf{v} = (2)(1) + (-3)(3) + (7)(1) = 2 - 9 + 7 = 0$$

$$\mathbf{u} \bullet \mathbf{w} = (2)(3) + (-3)(2) + (7)(0) = 6 - 6 + 0 = 0$$

$$\mathbf{v} \bullet \mathbf{w} = (1)(3) + (3)(2) + (1)(0) = 3 + 6 + 0 = 9.$$

Thus, **u** and **v** are orthogonal; **u** and **w** are orthogonal; but **v** and **w** are not orthogonal.   ■

Projections

In applications it is often necessary to express a given vector **u** in the form

$$\mathbf{u} = \mathbf{v} + \mathbf{w}$$

where

> **(a)** The vector **v** is parallel to a given nonzero vector **q**

> **(b)** The vector **w** is orthogonal to **q**

As an example, consider Figure 3 which illustrates a block on an inclined plane. In Figure 3, the downward directed force **u** is the weight of the block. As you can see, the force **u** has been expressed as the sum of two forces:

$$\mathbf{u} = \mathbf{v} + \mathbf{w}.$$

The force **v** is normal (perpendicular) to the inclined plane and (because of friction) tends to hold the block stationary on the plane. The force **w** is tangential to the inclined plane and tends to make the block slide downward. The relative magnitudes of **v** and **w** (along with the frictional characteristics of the block and the plane) determine whether the block remains stationary or slides.

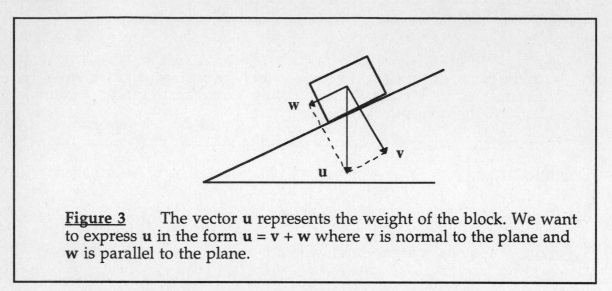

Figure 3   The vector **u** represents the weight of the block. We want to express **u** in the form **u** = **v** + **w** where **v** is normal to the plane and **w** is parallel to the plane.

The situation shown in Figure 3 is a special case of the general problem we want to discuss. The general problem is illustrated in Figure 4. As you see, Figure 4 shows a given vector **u** and a given direction defined by the vector **q**.

The vector **v** shown in Figure 4 is parallel to **q** and is called the *__projection of__ **u** __onto__ **q**__. For notation, we will write

$$\mathbf{v} = proj_{\mathbf{q}}(\mathbf{u}) \qquad (\text{"}\mathbf{v}\text{ is the projection of }\mathbf{u}\text{ onto }\mathbf{q}\text{"}).$$

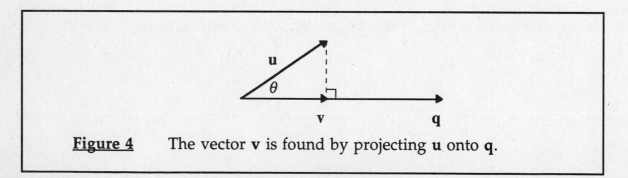

Figure 4   The vector **v** is found by projecting **u** onto **q**.

Calculating the projection

To calculate the projection **v** shown in Figure 4, we first write **v** as

$$\mathbf{v} = \|\mathbf{v}\| \left(\frac{\mathbf{q}}{\|\mathbf{q}\|} \right).$$

Next, from Figure 4, we see that $\|\mathbf{v}\| = \|\mathbf{u}\|\cos\theta$, and thus we have

$$(3) \qquad \mathbf{v} = \|\mathbf{u}\|\cos\theta\left(\frac{\mathbf{q}}{\|\mathbf{q}\|}\right).$$

By Theorem 1, $\mathbf{u} \bullet \mathbf{q} = \|\mathbf{u}\|\|\mathbf{q}\|\cos\theta$. Thus, in Equation (3), we can replace $\|\mathbf{u}\|\cos\theta$ by $\mathbf{u} \bullet \mathbf{q}/\|\mathbf{q}\|$. Making this replacement, we obtain a formula for the projection $\mathbf{v}$:

$$(4) \qquad \mathbf{v} = \frac{\mathbf{u} \bullet \mathbf{q}}{\|\mathbf{q}\|}\left(\frac{\mathbf{q}}{\|\mathbf{q}\|}\right).$$

Example 3   Let $\mathbf{u} = \begin{bmatrix} 2 \\ 1 \\ -3 \end{bmatrix}$ and $\mathbf{q} = \begin{bmatrix} 1 \\ 0 \\ 2 \end{bmatrix}$. Find $\mathbf{v} = proj_{\,\mathbf{q}}(\mathbf{u})$.

Solution: According to Equation (4),

$$\mathbf{v} = \frac{\mathbf{u} \bullet \mathbf{q}}{\|\mathbf{q}\|}\left(\frac{\mathbf{q}}{\|\mathbf{q}\|}\right) = \frac{-4}{\sqrt{5}}\left(\frac{\mathbf{q}}{\sqrt{5}}\right) = \frac{-4}{5}\begin{bmatrix} 1 \\ 0 \\ 2 \end{bmatrix} = \begin{bmatrix} -0.8 \\ 0 \\ -1.6 \end{bmatrix}.$$

   ∎

Depending on the relative magnitudes of $\mathbf{q}$ and $\mathbf{u}$, the projection of $\mathbf{u}$ onto $\mathbf{q}$ might be longer than $\mathbf{q}$ or shorter than $\mathbf{q}$, see Figure 5. Also, depending on the relative directions of $\mathbf{u}$ and $\mathbf{q}$, the projection of $\mathbf{u}$ onto $\mathbf{q}$ might be in the same direction as $\mathbf{q}$ or it might have the opposite direction, see Figure 5. Although formula (4) was derived from Figure 4, it is valid as well for the configurations shown in Figure 5 **(b)** - **(c)**.

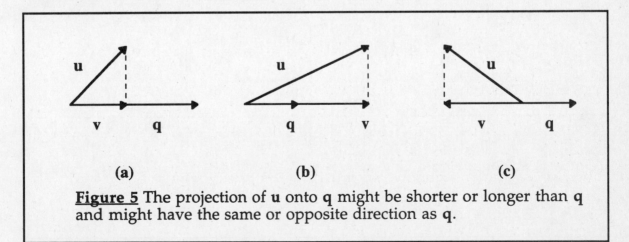

         (a)   **(b)**   **(c)**

Figure 5 The projection of $\mathbf{u}$ onto $\mathbf{q}$ might be shorter or longer than $\mathbf{q}$ and might have the same or opposite direction as $\mathbf{q}$.

Expressing u as the sum of two orthogonal vectors

We began our discussion of projections with the problem of expressing $\mathbf{u}$ in the form $\mathbf{u} = \mathbf{v} + \mathbf{w}$ where $\mathbf{v}$ is parallel to $\mathbf{q}$ and $\mathbf{w}$ is orthogonal to $\mathbf{q}$. Equation (4) gives us a formula for $\mathbf{v}$ and we see from Figure 6 that $\mathbf{w} = \mathbf{u} - \mathbf{v}$ is indeed orthogonal to $\mathbf{q}$. That is, the decomposition we sought has the form

$$\mathbf{u} = \mathbf{v} + \mathbf{w}$$

(5)

$$= proj_{\mathbf{q}}(\mathbf{u}) + (\mathbf{u} - proj_{\mathbf{q}}(\mathbf{u})).$$

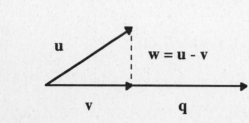

Figure 6   Here, $\mathbf{u} = \mathbf{v} + \mathbf{w}$. The vector $\mathbf{v}$ is the projection of $\mathbf{u}$ along $\mathbf{q}$. The vector $\mathbf{w} = \mathbf{u} - \mathbf{v}$ is orthogonal to $\mathbf{q}$, by construction.

Example 4   Let $\mathbf{u} = \begin{bmatrix} 4 \\ -2 \\ 3 \end{bmatrix}$ and $\mathbf{q} = \begin{bmatrix} 3 \\ 0 \\ 1 \end{bmatrix}$. Express $\mathbf{u}$ in the form $\mathbf{u} = \mathbf{v} + \mathbf{w}$ where $\mathbf{u}$ is parallel to $\mathbf{q}$ and $\mathbf{w}$ is orthogonal to $\mathbf{q}$; see Equation (5).

Solution: We first calculate the projection $\mathbf{v}$ as in Example 3:

$$\mathbf{v} = \frac{\mathbf{u} \bullet \mathbf{q}}{\|\mathbf{q}\|}\left(\frac{\mathbf{q}}{\|\mathbf{q}\|}\right) = \frac{15}{\sqrt{10}}\left(\frac{\mathbf{q}}{\sqrt{10}}\right) = \frac{15}{10}\begin{bmatrix} 3 \\ 0 \\ 1 \end{bmatrix} = \begin{bmatrix} 4.5 \\ 0 \\ 1.5 \end{bmatrix}.$$

Then, $\mathbf{w} = \mathbf{u} - \mathbf{v} = \begin{bmatrix} 4 \\ -2 \\ 3 \end{bmatrix} - \begin{bmatrix} 4.5 \\ 0 \\ 1.5 \end{bmatrix} = \begin{bmatrix} -0.5 \\ -2 \\ 1.5 \end{bmatrix}$. As we see, $\mathbf{v}$ is parallel to $\mathbf{q}$ since $\mathbf{v} =$ 1.5$\mathbf{q}$, $\mathbf{w}$ is orthogonal to $\mathbf{q}$, and $\mathbf{u} = \mathbf{v} + \mathbf{w}$. ∎

Exercises 3.6

1. In each of (a)-(d), calculate the dot product $\mathbf{u} \cdot \mathbf{v}$.

   (a) $\mathbf{u} = \begin{bmatrix} 1 \\ 3 \end{bmatrix}, \mathbf{v} = \begin{bmatrix} 4 \\ -2 \end{bmatrix}$.

   (b) $\mathbf{u} = \begin{bmatrix} 2 \\ 3 \end{bmatrix}, \mathbf{v} = \begin{bmatrix} -3 \\ 2 \end{bmatrix}$.

   (c) $\mathbf{u} = \begin{bmatrix} 1 \\ -2 \\ 1 \end{bmatrix}, \mathbf{v} = \begin{bmatrix} 3 \\ 1 \\ -2 \end{bmatrix}$.

   (d) $\mathbf{u} = 4\mathbf{i} + 2\mathbf{j} - 3\mathbf{k}, \mathbf{v} = -2\mathbf{i} + \mathbf{j} - 2\mathbf{k}$.

2. In each of (a)-(d), find $\cos\theta$, where θ is the angle between $\mathbf{u}$ and $\mathbf{v}$.

   (a) $\mathbf{u} = \begin{bmatrix} 3 \\ 1 \end{bmatrix}, \mathbf{v} = \begin{bmatrix} 2 \\ 5 \end{bmatrix}$.

   (b) $\mathbf{u} = \begin{bmatrix} 2 \\ 3 \end{bmatrix}, \mathbf{v} = \begin{bmatrix} -3 \\ 1 \end{bmatrix}$.

   (c) $\mathbf{u} = \begin{bmatrix} 1 \\ 2 \\ 1 \end{bmatrix}, \mathbf{v} = \begin{bmatrix} 2 \\ -1 \\ 1 \end{bmatrix}$.

   (d) $\mathbf{u} = 2\mathbf{i} - 3\mathbf{j} + \mathbf{k}, \mathbf{v} = \mathbf{i} - 2\mathbf{j} + 3\mathbf{k}$.

3. In each of (a)-(e), find θ (in radians), where θ is the angle between $\mathbf{u}$ and $\mathbf{v}$.

   (a) $\mathbf{u} = \begin{bmatrix} 3 \\ \sqrt{3} \end{bmatrix}, \mathbf{v} = \begin{bmatrix} 2 \\ 2\sqrt{3} \end{bmatrix}$.

   (b) $\mathbf{u} = \begin{bmatrix} \sqrt{3} \\ 1 \end{bmatrix}, \mathbf{v} = \begin{bmatrix} -3 \\ \sqrt{3} \end{bmatrix}$.

161

(c) $\mathbf{u} = \begin{bmatrix} 1 \\ 3 \\ -1 \end{bmatrix}$, $\mathbf{v} = \begin{bmatrix} 2 \\ -2 \\ -4 \end{bmatrix}$.

(d) $\mathbf{u} = \mathbf{i} + 2\mathbf{j} + \mathbf{k}$, $\mathbf{v} = 3\mathbf{i} + 6\mathbf{j} + 3\mathbf{k}$.

(e) $\mathbf{u} = -\mathbf{i} + \mathbf{j} - 2\mathbf{k}$, $\mathbf{v} = 2\mathbf{i} - 2\mathbf{j} + 4\mathbf{k}$.

4. In each of (a)-(f), there are at most two 3-dimensional vectors $\mathbf{u}$ that satisfy the given conditions. Determine these vectors $\mathbf{u}$.

   (a) $\mathbf{u} \cdot \mathbf{i} = 1$, $\mathbf{u} \cdot \mathbf{j} = 3$, $\mathbf{u} \cdot \mathbf{k} = 4$.

   (b) $\mathbf{u} \cdot \mathbf{i} = 0$, $\mathbf{u} \cdot \mathbf{j} = 0$, $\mathbf{u} \cdot \mathbf{k} = 4$.

   (c) $\mathbf{u} \cdot \mathbf{i} = 3$, $\mathbf{u} \cdot \mathbf{k} = 4$, $\|\mathbf{u}\| = 5$.

   (d) $\mathbf{u} \cdot \mathbf{i} = 12$, $\mathbf{u} \cdot \mathbf{k} = 3$, $\|\mathbf{u}\| = 13$.

   (e) $\mathbf{u} \cdot (\mathbf{i} + \mathbf{j}) = 2$, $\mathbf{u} \cdot (\mathbf{j} + \mathbf{k}) = 4$, $\mathbf{u} \cdot \mathbf{k} = 1$.

   (f) $\mathbf{u} \cdot (\mathbf{i} + \mathbf{j}) = 2$, $\mathbf{u} \cdot (\mathbf{j} + \mathbf{k}) = 3$, $\mathbf{u} \cdot \mathbf{k} = 2$.

5. In each of (a)-(d), $\mathbf{u} = \overrightarrow{OP}$ and $\mathbf{q} = \overrightarrow{OQ}$. If $\mathbf{w} = proj_{\mathbf{q}}\mathbf{u}$, then find the point R so that $\mathbf{w} = \overrightarrow{OR}$. Graph $\mathbf{u}$, $\mathbf{v}$, and $\mathbf{w}$.

   (a) $P = (2, 5)$, $Q = (6, 2)$.

   (b) $P = (7, 6)$, $Q = (4, 1)$.

   (c) $P = (-4, 2)$, $Q = (6, 2)$.

   (d) $P = (-2, 4)$, $Q = (4, 2)$.

6. In each of (a)-(d), for the given vectors $\mathbf{u}$ and $\mathbf{q}$ find vectors $\mathbf{u}_1$ and $\mathbf{u}_2$ such that (i) $\mathbf{u}_1 = proj_{\mathbf{q}}\mathbf{u}$, (ii) $\mathbf{u}_1$ and $\mathbf{u}_2$ are orthongonal, and (iii) $\mathbf{u} = \mathbf{u}_1 + \mathbf{u}_2$.

   (a) $\mathbf{u} = \begin{bmatrix} 7 \\ 3 \end{bmatrix}$, $\mathbf{q} = \begin{bmatrix} 1 \\ 1 \end{bmatrix}$.

(b)　$\mathbf{u} = \begin{bmatrix} 6 \\ 2 \end{bmatrix}$, $\mathbf{q} = \begin{bmatrix} 1 \\ -1 \end{bmatrix}$.

(c)　$\mathbf{u} = \begin{bmatrix} 6 \\ 4 \\ -2 \end{bmatrix}$, $\mathbf{q} = \begin{bmatrix} 1 \\ 2 \\ 1 \end{bmatrix}$.

(d)　$\mathbf{u} = 2\mathbf{i} + \mathbf{j} + 6\mathbf{k}$, $\mathbf{q} = \mathbf{i} + \mathbf{j} + \mathbf{k}$.

7.　Verify that $\mathbf{u} \cdot \mathbf{u} \geq 0$ for $\mathbf{u} = \begin{bmatrix} u_1 \\ u_2 \\ u_3 \end{bmatrix}$.

8.　Verify that $\mathbf{u} \cdot \mathbf{v} = \mathbf{v} \cdot \mathbf{u}$ for $\mathbf{u} = \begin{bmatrix} u_1 \\ u_2 \\ u_3 \end{bmatrix}$ and $\mathbf{v} = \begin{bmatrix} v_1 \\ v_2 \\ v_3 \end{bmatrix}$.

9.　If c is a scalar, then verify that $\mathbf{u} \cdot (c\mathbf{v}) = c(\mathbf{u} \cdot \mathbf{v})$ for $\mathbf{u} = \begin{bmatrix} u_1 \\ u_2 \\ u_3 \end{bmatrix}$ and

$\mathbf{v} = \begin{bmatrix} v_1 \\ v_2 \\ v_3 \end{bmatrix}$.

10.　Verify that $\mathbf{u} \cdot (\mathbf{v} + \mathbf{w}) = \mathbf{u} \cdot \mathbf{v} + \mathbf{u} \cdot \mathbf{w}$ for $\mathbf{u} = \begin{bmatrix} u_1 \\ u_2 \\ u_3 \end{bmatrix}$, $\mathbf{v} = \begin{bmatrix} v_1 \\ v_2 \\ v_3 \end{bmatrix}$,

and $\mathbf{w} = \begin{bmatrix} w_1 \\ w_2 \\ w_3 \end{bmatrix}$.

11.　Show that equation (1) simplifies to equation (2a) for $\mathbf{u} = \begin{bmatrix} u_1 \\ u_2 \end{bmatrix}$ and $\mathbf{v} = \begin{bmatrix} v_1 \\ v_2 \end{bmatrix}$.

3.7 THE CROSS PRODUCT

To introduce the cross product, let us consider the problem stated below and illustrated in Figure 1:

> **Problem:** Given nonzero vectors **u** and **v** in three space, find a vector **w** that is orthogonal to both **u** and **v**.

A solution to this problem is the *cross product vector*, the topic of this section.

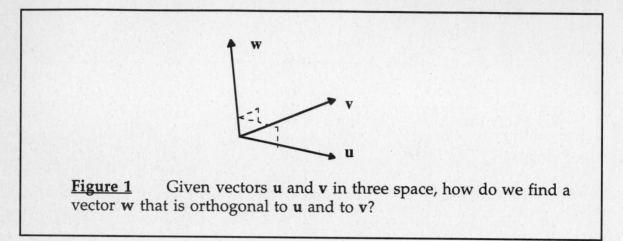

Figure 1   Given vectors **u** and **v** in three space, how do we find a vector **w** that is orthogonal to **u** and to **v**?

Derivation of the cross product

To solve the problem posed in Figure 1, let $\mathbf{u} = \begin{bmatrix} u_1 \\ u_2 \\ u_3 \end{bmatrix}$ and $\mathbf{v} = \begin{bmatrix} v_1 \\ v_2 \\ v_3 \end{bmatrix}$. We want a nonzero vector $\mathbf{w} = \begin{bmatrix} x \\ y \\ z \end{bmatrix}$ such that $\mathbf{w} \bullet \mathbf{u} = 0$ and $\mathbf{w} \bullet \mathbf{v} = 0$. In terms of the components of **u**, **v**, and **w**, the orthogonality conditions lead to a system of two equations in three unknowns:

$$u_1 x + u_2 y + u_3 z = 0$$

$$v_1 x + v_2 y + v_3 z = 0$$

The system above is homogeneous and has more unknowns than equations. Thus, see Section 1.8, we know the system always has nonzero solutions. In the exercises, we ask you to show that one nonzero solution is

(1) $x = u_2 v_3 - u_3 v_2$, $y = u_3 v_1 - u_1 v_3$, $z = u_1 v_2 - u_2 v_1$.

The vector **w** having components (1) is called the ***cross product*** of **u** and **v**.

The definition of the cross product

Drawing on the derivation above, we are led to the following definition

---

DEFINITION 1   Let $\mathbf{u} = \begin{bmatrix} u_1 \\ u_2 \\ u_3 \end{bmatrix}$ and $\mathbf{v} = \begin{bmatrix} v_1 \\ v_2 \\ v_3 \end{bmatrix}$ be vectors in three space.

The _**cross product**_ of $\mathbf{u}$ and $\mathbf{v}$, denoted $\mathbf{u} \times \mathbf{v}$, is the vector given by

$$\mathbf{u} \times \mathbf{v} = \begin{bmatrix} u_2 v_3 - u_3 v_2 \\ u_3 v_1 - u_1 v_3 \\ u_1 v_2 - u_2 v_1 \end{bmatrix}.$$

The next example illustrates a typical cross product calculation.

Example 1   Find $\mathbf{u} \times \mathbf{v}$ for $\mathbf{u} = \begin{bmatrix} 2 \\ 1 \\ 3 \end{bmatrix}$ and $\mathbf{v} = \begin{bmatrix} 1 \\ -3 \\ 4 \end{bmatrix}$. Verify that the cross product vector is orthogonal to both $\mathbf{u}$ and $\mathbf{v}$.

Solution: Using Definition 1, we find the cross product vector

$$\mathbf{u} \times \mathbf{v} = \begin{bmatrix} (1)(4) - (3)(-3) \\ (3)(1) - (2)(4) \\ (2)(-3) - (1)(1) \end{bmatrix} = \begin{bmatrix} 13 \\ -5 \\ -7 \end{bmatrix}.$$

A calculation shows that $(\mathbf{u} \times \mathbf{v}) \bullet \mathbf{u} = 0$ and $(\mathbf{u} \times \mathbf{v}) \bullet \mathbf{v} = 0$. ■

Using determinants to remember the form of the cross product

We can use a "symbolic determinant" to help us remember the cross product:

$$(2) \qquad \mathbf{u} \times \mathbf{v} = \begin{vmatrix} \mathbf{i} & \mathbf{j} & \mathbf{k} \\ u_1 & u_2 & u_3 \\ v_1 & v_2 & v_3 \end{vmatrix} = (u_2 v_3 - u_3 v_2)\mathbf{i} - (u_1 v_3 - u_3 v_1)\mathbf{j} + (u_1 v_2 - u_2 v_1)\mathbf{k}.$$

Example 2   Use the symbolic determinant (2) to calculate $\mathbf{u} \times \mathbf{v}$ and $\mathbf{v} \times \mathbf{u}$, where $\mathbf{u} = \begin{bmatrix} 1 \\ 1 \\ -1 \end{bmatrix}$ and $\mathbf{v} = \begin{bmatrix} 2 \\ -1 \\ 3 \end{bmatrix}$.

Solution: Using (2), we obtain

$$\mathbf{u} \times \mathbf{v} = \begin{vmatrix} \mathbf{i} & \mathbf{j} & \mathbf{k} \\ 1 & 1 & -1 \\ 2 & -1 & 3 \end{vmatrix} = (3-1)\mathbf{i} - (3+2)\mathbf{j} + (-1-2)\mathbf{k} = 2\mathbf{i} - 5\mathbf{j} - 3\mathbf{k} = \begin{bmatrix} 2 \\ -5 \\ -3 \end{bmatrix}.$$

To calculate $\mathbf{v} \times \mathbf{u}$ we switch the rows in the symbolic determinant for $\mathbf{u} \times \mathbf{v}$:

$$\mathbf{v} \times \mathbf{u} = \begin{vmatrix} \mathbf{i} & \mathbf{j} & \mathbf{k} \\ 2 & -1 & 3 \\ 1 & 1 & -1 \end{vmatrix}.$$

From the properties of determinants, we know that the determinant changes sign when two rows are interchanged. Thus, even without expanding the determinant for $\mathbf{v} \times \mathbf{u}$, we know that $\mathbf{v} \times \mathbf{u} = -\mathbf{u} \times \mathbf{v} = \begin{bmatrix} -2 \\ 5 \\ 3 \end{bmatrix}.$ ∎

Algebraic properties of the cross product

We have to be careful with cross product calculations since properties that we intuitively expect to hold may not be true. For instance, we saw in Example 2 that the expected equality, $\mathbf{u} \times \mathbf{v} = \mathbf{v} \times \mathbf{u}$, may not be true.

A summary of several important algebraic properties of the cross product is given in Theorem 1.

THEOREM 1   Let $\mathbf{u}$, $\mathbf{v}$, and $\mathbf{w}$ be three-dimensional vectors and let c be a scalar. Then

(a) $\mathbf{u} \times \mathbf{v} = -\mathbf{v} \times \mathbf{u}$ (b) $\mathbf{u} \times \mathbf{u} = 0$ (c) $(c\mathbf{u}) \times \mathbf{v} = \mathbf{u} \times (c\mathbf{v}) = c(\mathbf{u} \times \mathbf{v})$

(d) $\mathbf{u} \times (\mathbf{v} + \mathbf{w}) = \mathbf{u} \times \mathbf{v} + \mathbf{u} \times \mathbf{w}$ (e) $\mathbf{u} \bullet (\mathbf{v} \times \mathbf{w}) = (\mathbf{u} \times \mathbf{v}) \bullet \mathbf{w}$

Proof: The proof of Theorem 1 follows from the definition of the cross product. ∎

It is worth noting that the cross product operation is not associative. That is, in general, $(\mathbf{u} \times \mathbf{v}) \times \mathbf{w}$ is not equal to $\mathbf{u} \times (\mathbf{v} \times \mathbf{w})$, see the exercises for an example.

The right-hand rule

By Theorem 1, $\mathbf{u} \times \mathbf{v} = -\mathbf{v} \times \mathbf{u}$. Therefore, both of the vectors $\mathbf{u} \times \mathbf{v}$ and $\mathbf{v} \times \mathbf{u}$ are perpendicular to $\mathbf{u}$ and to $\mathbf{v}$.

The orientation of $\mathbf{u} \times \mathbf{v}$ can be found by using the *__right-hand rule__*:

> Let $\mathbf{u}$, $\mathbf{v}$, and $\mathbf{u} \times \mathbf{v}$ have a common initial point A. Imagine placing your right hand near A with your fingers pointing along $\mathbf{u}$. Let your fingers curl from $\mathbf{u}$ towards $\mathbf{v}$. Your thumb points in the direction of $\mathbf{u} \times \mathbf{v}$.

Cross products of the basic unit vectors illustrate the right-hand rule. The first line of cross products can be found using Definition 1 or the determinant (2). The other two lines follow from Theorem 1.

$$
\begin{array}{lll}
\mathbf{i} \times \mathbf{j} = \mathbf{k} & \mathbf{j} \times \mathbf{k} = \mathbf{i} & \mathbf{k} \times \mathbf{i} = \mathbf{j} \\
\mathbf{j} \times \mathbf{i} = -\mathbf{k} & \mathbf{k} \times \mathbf{j} = -\mathbf{i} & \mathbf{i} \times \mathbf{k} = -\mathbf{j} \\
\mathbf{i} \times \mathbf{i} = 0 & \mathbf{j} \times \mathbf{j} = 0 & \mathbf{k} \times \mathbf{k} = 0
\end{array}
$$

(3)

When $\mathbf{u}$ and $\mathbf{v}$ are given in terms of the basic unit vectors, we can use Theorem 1 and Equation (3) to calculate $\mathbf{u} \times \mathbf{v}$. The next example illustrates the idea.

Example 3   Let $\mathbf{u} = 3\mathbf{i} + 2\mathbf{k}$ and let $\mathbf{v} = 4\mathbf{i} + \mathbf{j} + 5\mathbf{k}$. Calculate $\mathbf{u} \times \mathbf{v}$.

Solution: We have

$$\mathbf{u} \times \mathbf{v} = (3\mathbf{i} + 2\mathbf{k}) \times (4\mathbf{i} + \mathbf{j} + 5\mathbf{k})$$

$$= 12\mathbf{i} \times \mathbf{i} + 3\mathbf{i} \times \mathbf{j} + 15\mathbf{i} \times \mathbf{k} + 8\mathbf{k} \times \mathbf{i} + 2\mathbf{k} \times \mathbf{j} + 10\mathbf{k} \times \mathbf{k}$$

$$= 120 + 3\mathbf{k} - 15\mathbf{j} + 8\mathbf{j} - 2\mathbf{i} + 100$$

$$= -2\mathbf{i} - 7\mathbf{j} + 3\mathbf{k}$$

∎

Geometric properties of the cross product

The right-hand rule gives us the direction of u×v. Theorem 2 tells us the magnitude of u×v.

THEOREM 2   Let **u** and **v** be nonzero three-dimensional vectors and let θ be the angle between **u** and **v**. The length of **u**×**v** is given by

$$\|\mathbf{u} \times \mathbf{v}\| = \|\mathbf{u}\| \, \|\mathbf{v}\| \sin\theta.$$

We leave the proof of Theorem 2 as an exercise. Note that Theorem 2 can be used as a test of whether or not nonzero vectors **u** and **v** are parallel; see Theorem 4 at the end of this section.

An interesting application of Theorem 2 involves using **u**×**v** to find the area of a parallelogram. In particular, let A, B, C, and D be four points in space that lie on the same plane. Let $ABCD$ denote the parallelogram determined by these four points, see Figure 2.

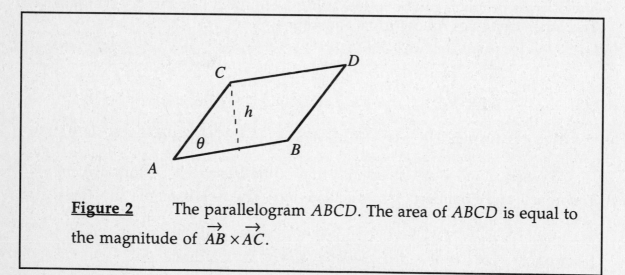

Figure 2   The parallelogram $ABCD$. The area of $ABCD$ is equal to the magnitude of $\overrightarrow{AB} \times \overrightarrow{AC}$.

The following result can be established with the help of Figure 2.

THEOREM 3   Let A, B, C, and D be points in three space, that lie on a plane. Let $ABCD$ be the parallelogram determined by these points, where A and D are opposite vertices. Then the area of $ABCD$ is equal to the length of the cross product vector $\overrightarrow{AB} \times \overrightarrow{AC}$.

Proof: We can verify Theorem 3 by appealing to Figure 2:

$$\text{Area} = \left\| \overrightarrow{AB} \right\| h \qquad\qquad (\text{area} = \text{base} \times \text{height})$$

$$= \left\| \overrightarrow{AB} \right\| \left\| \overrightarrow{AC} \right\| \sin\theta \qquad\qquad (\text{since } h = \left\| \overrightarrow{AC} \right\| \sin\theta)$$

$$= \left\| \overrightarrow{AB} \times \overrightarrow{AC} \right\| \qquad\qquad (\text{by Theorem 2})$$

Example 4 Use Theorem 3 to find the area of the triangle ABC where the vertices are $A = (1, 2, 2)$, $B = (3, 1, 4)$, and $C = (5, 2, 1)$.

Solution: From geometry, we know that the triangle ABC has half the area of the parallelogram $ABCD$, see Figure 2. So, as in Figure 2, let $\mathbf{v} = \overrightarrow{AB}$ and let $\mathbf{u} = \overrightarrow{AC}$. Forming $\mathbf{u} \times \mathbf{v}$ we find

$$\mathbf{u} \times \mathbf{v} = \begin{vmatrix} \mathbf{i} & \mathbf{j} & \mathbf{k} \\ 4 & 0 & -1 \\ 2 & -1 & 2 \end{vmatrix} = -\mathbf{i} - 10\mathbf{j} - 4\mathbf{k}.$$

Calculating the magnitude of $\mathbf{u} \times \mathbf{v}$, we obtain $\sqrt{117}$. Thus, the area of the triangle ABC is equal to $\sqrt{117}/2$. ∎

Triple products

The products $\mathbf{u} \bullet (\mathbf{v} \times \mathbf{w})$ and $\mathbf{u} \times (\mathbf{v} \times \mathbf{w})$ are called *__triple products__*. The triple product $\mathbf{u} \bullet (\mathbf{v} \times \mathbf{w})$ is known as the *__scalar triple product__* while $\mathbf{u} \times (\mathbf{v} \times \mathbf{w})$ is called the *__vector triple product__*.

The magnitude of the scalar triple product has an interesting geometric interpretation. Suppose PQ, PR, and PS are adjacent edges of a parallelepiped, as in Figure 3. Let $\mathbf{u} = \overrightarrow{PQ}$, $\mathbf{v} = \overrightarrow{PR}$, and $\mathbf{w} = \overrightarrow{PS}$. In the exercises we ask you to verify that the volume, V, of the parallelepiped is given by

(4) $$V = \left| \mathbf{u} \bullet (\mathbf{v} \times \mathbf{w}) \right|.$$

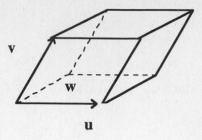

Figure 3   The volume of the parallelepiped is $|\mathbf{u}\bullet(\mathbf{v}\times\mathbf{w})|$.

Tests for collinearity and coplanarity

Let A be the area of a parallelogram with adjacent edges $\mathbf{u} = \overrightarrow{AB}$ and $\mathbf{v} = \overrightarrow{AC}$. Then, see Figure 2, $A = 0$ if and only if $\mathbf{u}$ and $\mathbf{v}$ are collinear. Similarly, let V be the volume of the parallelepiped in Figure 3. We see that $V = 0$ if and only if the vectors $\mathbf{u}$, $\mathbf{v}$, and $\mathbf{w}$ are coplanar.

Combining $A = 0$ with Theorem 3 and $V = 0$ with Equation (4), we obtain the following useful test.

THEOREM 4   (a) Let $\mathbf{u}$ and $\mathbf{v}$ be nonzero three-dimensional vectors. Then $\mathbf{u}$ and $\mathbf{v}$ are collinear if and only if $\mathbf{u}\times\mathbf{v} = \mathbf{0}$.

(b) Let $\mathbf{u}$, $\mathbf{v}$, and $\mathbf{w}$ be nonzero three-dimensional vectors. Then $\mathbf{u}$, $\mathbf{v}$, and $\mathbf{w}$ are coplanar if and only if $\mathbf{u}\bullet(\mathbf{v}\times\mathbf{w}) = 0$.

170

Exercises 3.7

1. In each of (a)-(f), calculate the cross product, $\mathbf{u} \times \mathbf{v}$.

    (a) $\mathbf{u} = \begin{bmatrix} 1 \\ 3 \\ 0 \end{bmatrix}, \mathbf{v} = \begin{bmatrix} 4 \\ 1 \\ 0 \end{bmatrix}.$

    (b) $\mathbf{u} = \begin{bmatrix} 4 \\ 0 \\ 1 \end{bmatrix}, \mathbf{v} = \begin{bmatrix} 2 \\ 2 \\ 0 \end{bmatrix}.$

    (c) $\mathbf{u} = \begin{bmatrix} 2 \\ 1 \\ 1 \end{bmatrix}, \mathbf{v} = \begin{bmatrix} 1 \\ 8 \\ 1 \end{bmatrix}.$

    (d) $\mathbf{u} = \mathbf{i} + \mathbf{j} + 3\mathbf{k}, \mathbf{v} = 2\mathbf{i} + 2\mathbf{j} + 6\mathbf{k}.$

    (e) $\mathbf{u} = \mathbf{i} + 3\mathbf{j}, \mathbf{v} = 2\mathbf{i} + \mathbf{j} + \mathbf{k}.$

    (f) $\mathbf{u} = 2\mathbf{i} + \mathbf{j} - \mathbf{k}, \mathbf{v} = \mathbf{i} + 3\mathbf{j} + 2\mathbf{k}.$

2. In each of (a)-(d), find a vector $\mathbf{w}$ such that $\mathbf{u} \cdot \mathbf{w} = 0$ and $\mathbf{v} \cdot \mathbf{w} = 0$.

    (a) $\mathbf{u} = \begin{bmatrix} 3 \\ 1 \\ 2 \end{bmatrix}, \mathbf{v} = \begin{bmatrix} 1 \\ 1 \\ 1 \end{bmatrix}.$

    (b) $\mathbf{u} = \begin{bmatrix} 2 \\ 0 \\ 1 \end{bmatrix}, \mathbf{v} = \begin{bmatrix} 1 \\ 2 \\ 3 \end{bmatrix}.$

    (c) $\mathbf{u} = \mathbf{i} + \mathbf{j}, \mathbf{v} = \mathbf{i} + \mathbf{k}.$

    (d) $\mathbf{u} = \mathbf{i} - 2\mathbf{k}, \mathbf{v} = \mathbf{j} + 3\mathbf{k}.$

3. In each of (a) and (b), find a vector $\mathbf{w}$ that is perpendicular to the plane containing the given points A, B, and C.

    (a) $A = (-1, 1, 2), B = (2, 1, -1), C = (0, -2, 4).$

(b) $A = (1, 0, 0)$, $B = (0, 1, 0)$, $C = (2, 3, 1)$.

4. In each of (a) and (b), two sides of a parallelogram coincide with the position vectors for $\mathbf{u}$ and $\mathbf{v}$. Determine the area of the parallelogram.

(a) $\mathbf{u} = \begin{bmatrix} 1 \\ 2 \\ 4 \end{bmatrix}$, $\mathbf{v} = \begin{bmatrix} 0 \\ 2 \\ 1 \end{bmatrix}$.

(b) $\mathbf{u} = \begin{bmatrix} 4 \\ -1 \\ 2 \end{bmatrix}$, $\mathbf{v} = \begin{bmatrix} 0 \\ 1 \\ 2 \end{bmatrix}$.

5. In (a) and (b), find the area of the triangle having the given points as vertices.

(a) $A = (0, 0, 0)$, $B = (2, 3, 4)$, $C = (1, -1, 2)$.

(b) $A = (5, -1, 1)$, $B = (4, 3, 0)$, $C = (0, 1, 2)$.

6. In (a) and (b), three edges of a parallelepiped coincide with the position vectors for $\mathbf{u}$, $\mathbf{v}$, and $\mathbf{w}$. Determine the volume of the parallelepiped.

(a) $\mathbf{u} = \begin{bmatrix} 0 \\ 1 \\ 3 \end{bmatrix}$, $\mathbf{v} = \begin{bmatrix} -2 \\ 1 \\ 0 \end{bmatrix}$, $\mathbf{w} = \begin{bmatrix} 0 \\ 4 \\ 1 \end{bmatrix}$.

(b) $\mathbf{u} = \begin{bmatrix} 2 \\ 2 \\ 0 \end{bmatrix}$, $\mathbf{v} = \begin{bmatrix} 1 \\ 4 \\ 0 \end{bmatrix}$, $\mathbf{w} = \begin{bmatrix} 1 \\ 2 \\ 4 \end{bmatrix}$.

7. In (a) and (b), determine if the three given vectors $\mathbf{u}$, $\mathbf{v}$, and $\mathbf{w}$ are coplanar.

(a) $\mathbf{u} = \begin{bmatrix} 1 \\ -1 \\ 0 \end{bmatrix}$, $\mathbf{v} = \begin{bmatrix} 2 \\ 0 \\ 1 \end{bmatrix}$, $\mathbf{w} = \begin{bmatrix} 0 \\ 2 \\ 1 \end{bmatrix}$.

(b) $\quad \mathbf{u} = \begin{bmatrix} 2 \\ -2 \\ 1 \end{bmatrix}, \mathbf{v} = \begin{bmatrix} 0 \\ 1 \\ -1 \end{bmatrix}, \mathbf{w} = \begin{bmatrix} -1 \\ 0 \\ 3 \end{bmatrix}.$

8. Verify that $x = u_2 v_3 - u_3 v_2$, $y = u_3 v_1 - u_1 v_3$, $z = u_1 v_2 - u_2 v_1$ is a solution for the system of equations

$$\begin{array}{ccccccc} u_1 x & + & u_2 y & + & u_3 z & = & 0 \\ v_1 x & + & v_2 y & + & v_3 z & = & 0 \end{array}$$

9. Show that $(\mathbf{i} \times \mathbf{i}) \times \mathbf{j} \neq \mathbf{i} \times (\mathbf{i} \times \mathbf{j})$.

10. Let $\mathbf{u} = \begin{bmatrix} u_1 \\ u_2 \\ u_3 \end{bmatrix}$ and $\mathbf{v} = \begin{bmatrix} v_1 \\ v_2 \\ v_3 \end{bmatrix}$.

   (a) Verify that $\|\mathbf{u} \times \mathbf{v}\|^2 = \|\mathbf{u}\|^2 \|\mathbf{v}\|^2 - (\mathbf{u} \cdot \mathbf{v})^2$.

   (b) Use the equation in (a) to show $\|\mathbf{u} \times \mathbf{v}\| = \|\mathbf{u}\| \|\mathbf{v}\| \sin\theta$, where θ is the angle between $\mathbf{u}$ and $\mathbf{v}$. [HINT: Recall that $\mathbf{u} \cdot \mathbf{v} = \|\mathbf{u}\| \|\mathbf{v}\| \cos\theta$.]

11. Suppose PQ, PR, and PS are adjacent edges of a rectangular parallelepiped, as in Figure 3. Let $\mathbf{u} = \overrightarrow{PQ}$, $\mathbf{v} = \overrightarrow{PR}$, $\mathbf{w} = \overrightarrow{PS}$. Show that the volume, V, of the parallelepiped is given by $V = |\mathbf{u} \cdot (\mathbf{v} \times \mathbf{w})|$. [HINT: Recall that $|\mathbf{u} \cdot (\mathbf{v} \times \mathbf{w})| = \|\mathbf{u}\| \|\mathbf{v} \times \mathbf{w}\| |\cos\theta|$, where θ is the angle between $\mathbf{u}$ and $\mathbf{v} \times \mathbf{w}$. Show that $\|\mathbf{u}\| |\cos\theta|$ is the height of the parallelepiped and recall that $\|\mathbf{v} \times \mathbf{w}\|$ is the area of the parallelogram that forms the base of the parallelepiped.]

4.1 THE EIGENVALUE PROBLEM

In this chapter we study the _**eigenvalue problem**_. Eigenvalues are very important in applied mathematics arising in areas such as differential equations, control theory, optimization, and so forth.

We use Figure 1 to introduce eigenvalues. Let A be an $(n \times n)$ matrix and let **u** be an n-dimensional vector. When we form the product A**u**, it usually happens that the vectors **u** and A**u** point in different directions, see Figure 1(a).

However, there are certain special vectors **x** such that **x** and A**x** are _parallel_, see Figure 1(b). As we shall see, such a special vector **x** is called an _eigenvector_.

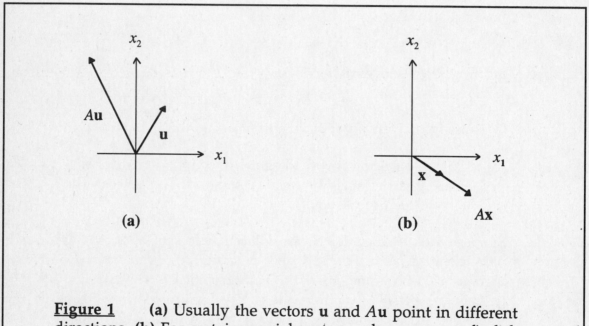

Figure 1 **(a)** Usually the vectors **u** and A**u** point in different directions. **(b)** For certain special vectors **x**, however, we find that **x** and A**x** are parallel.

The definition of eigenvalues and eigenvectors

In Figure 1(b) A**x** and **x** are parallel vectors. As we know, saying A**x** is parallel to **x** means that A**x** is a scalar multiple of **x**; that is, A**x** $= \lambda$**x** for some scalar λ.

> **THE EIGENVALUE PROBLEM**   Let A be a given $(n \times n)$ matrix. Find   all numbers λ such that Equation (1) has a nonzero solution x:
>
> (1) $$Ax = \lambda x.$$
>
> Such a scalar λ is called an *eigenvalue* of A. Any nonzero vector x satisfying Equation (1) is called an *eigenvector* corresponding to λ.

We restrict x to be nonzero in Equation (1) so as to rule out trivial cases. If we allow $x = 0$ in Equation (1), then every real number λ would be an eigenvalue.

Example 1   Let $A = \begin{bmatrix} 3 & 5 \\ 1 & -1 \end{bmatrix}$, $\lambda = 4$, and $x = \begin{bmatrix} 5 \\ 1 \end{bmatrix}$. Verify that $\lambda = 4$ is an eigenvalue of A with a corresponding eigenvector x. That is, show that the nonzero vector x together with the number $\lambda = 4$ solves Equation (1).

Solution: By direct multiplication, we have:

$$Ax = \begin{bmatrix} 3 & 5 \\ 1 & -1 \end{bmatrix}\begin{bmatrix} 5 \\ 1 \end{bmatrix} = \begin{bmatrix} 20 \\ 4 \end{bmatrix}$$

$$= 4\begin{bmatrix} 5 \\ 1 \end{bmatrix}$$

$$= 4x$$

Thus, since $Ax = 4x$, we see from Equation (1) that $\lambda = 4$ is an eigenvalue, with a corresponding eigenvector x. ∎

An eigenvalue has infinitely many eigenvectors

Note that Equation (1), $Ax = \lambda x$, is a system of n equations, but with $n + 1$ unknowns. The unknowns are the n components of x (namely $x_1, x_2, \ldots, x_n$) and the value λ. Also note that the system (1) is nonlinear since the unknown λ multiplies each of the unknowns $x_1, x_2, \ldots, x_n$. The nonlinear system (1) is unlike any other system of equations we have studied in this book and we now consider how we might solve it.

Let us first rewrite $A\mathbf{x} = \lambda\mathbf{x}$ by moving $\lambda\mathbf{x}$ to the left-hand side:

$$A\mathbf{x} - \lambda\mathbf{x} = \mathbf{0}.$$

Factoring the vector $\mathbf{x}$ to the right, we next obtain an equivalent version of Equation (1):

(2) $$(A - \lambda I)\mathbf{x} = \mathbf{0}.$$

From Equation (2) we see that if $\mathbf{x}$ is an eigenvector corresponding to λ, then so is $\mathbf{w} = c\mathbf{x}$, c a nonzero scalar.

λ is an eigenvalue of A if and only if $A - \lambda I$ is not invertible

We now turn to the problem of solving Equation (2). If the matrix $A - \lambda I$ happens to be invertible, then we can multiply both sides of Equation (2) by the inverse to obtain:

$$\mathbf{x} = (A - \lambda I)^{-1}\mathbf{0}.$$

So, if $A - \lambda I$ is invertible, then the only vector $\mathbf{x}$ satisfying Equation (2) is the vector $\mathbf{x} = \mathbf{0}$.

But, $\mathbf{x} = \mathbf{0}$ is not a permissible solution for the eigenvalue problem. Therefore, we have to conclude: if λ is an eigenvalue of A, then $A - \lambda I$ *must not* have an inverse.

THEOREM 1   Let A be an $(n \times n)$ matrix. Then λ is an eigenvalue of A if and only if $A - \lambda I$ is not invertible.

Proof: We prove Theorem 1 by calling on Theorem 5 of Section 2.8 which implies that the equation $(A - \lambda I)\mathbf{x} = \mathbf{0}$ has nontrivial solutions if and only if $A - \lambda I$ is not invertible.   ∎

Because of Theorem 1, we see that the eigenvalue problem can be solved using the following a two-step process.

Two-step process for solving the eigenvalue problem

Step 1:   Find all numbers λ such that $A - \lambda I$ is not invertible.

Step 2:   Given a number λ such that $A - \lambda I$ is not invertible, find all nonzero vectors $\mathbf{x}$ such that $(A - \lambda I)\mathbf{x} = \mathbf{0}$.

In the next subsection we see how to carry out Step 1—finding the eigenvalues of A. Once we know the eigenvalues, it is easy to carry out Step 2—finding the eigenvectors. In particular, once we have the eigenvalues from Step 1, we can use Gauss-Jordan elimination to solve $(A - \lambda I)\mathbf{x} = \mathbf{0}$ for the eigenvectors.

Using determinants to find eigenvalues

Theorem 1 suggests a method for finding eigenvalues. In particular, there is a simple test for whether or not $A - \lambda I$ is invertible. For a given number λ we know from Theorem 4, Section 2.8:

$$A - \lambda I \text{ is not invertible} \Leftrightarrow \det(A - \lambda I) = 0.$$

Therefore, λ is an eigenvalue of A if and only if

(3) $$\det(A - \lambda I) = 0.$$

So, if we want to find all the eigenvalues of a matrix A, we can do so by solving Equation (3)—finding all values λ such that $\det(A - \lambda I) = 0$.

Example 2   Let $A = \begin{bmatrix} 5 & -2 \\ 6 & -2 \end{bmatrix}$. Solve Equation (3) to find the eigenvalues of A.

Solution: For this matrix A, Equation (3) is given by

$$\begin{vmatrix} 5 - \lambda & -2 \\ 6 & -2 - \lambda \end{vmatrix} = 0,$$

or

$$(5 - \lambda)(-2 - \lambda) + 12 = 0.$$

Expanding, we obtain a quadratic equation:

$$\lambda^2 - 3\lambda + 2 = 0,$$

or

$$(\lambda - 2)(\lambda - 1) = 0.$$

Therefore, the solutions of $\det(A - \lambda I) = 0$ are $\lambda = 2$ and $\lambda = 1$. In other words, the eigenvalues of A are $\lambda = 2$ and $\lambda = 1$. ■

The characteristic polynomial

We now formalize the process illustrated in Example 2—using determinants to find eigenvalues. We begin with Equation (3) from the last subsection:

$$\det(A - \lambda I) = 0.$$

It can be shown that $\det(A - \lambda I)$ is a *polynomial* of degree n whenever A is an $(n \times n)$ matrix. The polynomial $p(\lambda) = \det(A - \lambda I)$ is called the ***characteristic polynomial*** and the equation $p(\lambda) = 0$ is called the ***characteristic equation***. We summarize our discussion in Theorem 2.

THEOREM 2　　　Let A be an $(n \times n)$ matrix and let $p(\lambda) = 0$ be　the characteristic equation for A. The eigenvalues of A are precisely the roots of $p(\lambda) = 0$.

Theorem 2 gives us a simple method for finding all the eigenvalues. The next example illustrates Theorem 2.

Example 3　　Let $A = \begin{bmatrix} 2 & 1 \\ 1 & 2 \end{bmatrix}$. Find the characteristic polynomial, $p(\lambda)$. Solve the characteristic equation $p(\lambda) = 0$ to find the eigenvalues of A.

Solution:　　The characteristic polynomial is:

$$p(\lambda) = \det\begin{bmatrix} 2 - \lambda & 1 \\ 1 & 2 - \lambda \end{bmatrix}$$

$$= (2 - \lambda)(2 - \lambda) - 1$$

$$= \lambda^2 - 4\lambda + 3$$

Having the characteristic polynomial, we obtain the characteristic equation

$$\lambda^2 - 4\lambda + 3 = 0$$

or, after factoring,

$$(\lambda - 3)(\lambda - 1) = 0.$$

Therefore, the eigenvalues are $\lambda = 3$ and $\lambda = 1$. ∎

Finding Eigenvectors

Once we know the eigenvalues for A, we find the corresponding eigenvectors by solving $(A - \lambda I)\mathbf{x} = \mathbf{0}$. The *nonzero* solutions of this homogeneous equation are the eigenvectors corresponding to λ.

Example 4   Find the eigenvalues and eigenvectors for $A = \begin{bmatrix} 2 & 5 \\ 6 & 1 \end{bmatrix}$.

Solution: The characteristic polynomial is

$$\det(A - \lambda I) = \begin{vmatrix} 2 - \lambda & 5 \\ 6 & 1 - \lambda \end{vmatrix}$$

$$= \lambda^2 - 3\lambda - 28$$

$$= (\lambda - 7)(\lambda + 4)$$

Thus, the eigenvalues are $\lambda = 7$ and $\lambda = -4$.

Eigenvectors corresponding to $\lambda = 7$:

To find the eigenvectors corresponding to $\lambda = 7$, we need the nonzero solutions of $(A - 7I)\mathbf{x} = \mathbf{0}$. Now, the matrix $A - 7I$ is given by $\begin{bmatrix} -5 & 5 \\ 6 & -6 \end{bmatrix}$. Forming the augmented matrix for the system $(A - 7I)\mathbf{x} = \mathbf{0}$ we obtain:

$$[A - 7I \mid \mathbf{0}] = \begin{bmatrix} -5 & 5 & 0 \\ 6 & -6 & 0 \end{bmatrix} \xrightarrow[R_2 - 6R_1]{(-1/5)R_1} \begin{bmatrix} 1 & 1 & 0 \\ 0 & 0 & 0 \end{bmatrix}.$$

Therefore, the system $(A - 7I)\mathbf{x} = \mathbf{0}$ reduces to

$$x_1 + x_2 = 0.$$

Thus, the eigenvectors are all of the form

$$\mathbf{x} = \begin{bmatrix} x_1 \\ x_2 \end{bmatrix} = \begin{bmatrix} -x_2 \\ x_2 \end{bmatrix} = x_2 \begin{bmatrix} -1 \\ 1 \end{bmatrix}, \quad x_2 \neq 0.$$

There are infinitely many eigenvectors corresponding to $\lambda = 7$. By making different choices for x_2, we generate different eigenvectors; for example, the vectors below are all eigenvectors corresponding to $\lambda = 7$:

$$\mathbf{x} = \begin{bmatrix} -1 \\ 1 \end{bmatrix}, \quad \mathbf{x} = \begin{bmatrix} -8 \\ 8 \end{bmatrix}, \quad \mathbf{x} = \begin{bmatrix} 13 \\ -13 \end{bmatrix}, \quad \mathbf{x} = \begin{bmatrix} 4.9 \\ -4.9 \end{bmatrix}.$$

Eigenvectors corresponding to λ = -4:

We next find the eigenvectors for the other eigenvalue, $\lambda = -4$. The eigenvectors corresponding to $\lambda = -4$ are nonzero solutions of $(A + 4I)\mathbf{x} = \mathbf{0}$. The augmented matrix for this system is

$$[A + 4I \mid \mathbf{0}] = \begin{bmatrix} 6 & 5 & 0 \\ 6 & 5 & 0 \end{bmatrix}.$$

Solving the system, we find $x_1 = -(5/6)x_2$. Therefore, the eigenvectors corresponding to $\lambda = -4$ are given by:

$$\mathbf{x} = \begin{bmatrix} x_1 \\ x_2 \end{bmatrix} = \begin{bmatrix} (-5/6)x_2 \\ x_2 \end{bmatrix} = x_2 \begin{bmatrix} -5/6 \\ 1 \end{bmatrix}, \quad x_2 \neq 0.$$

For instance, choosing $x_2 = -6$, we obtain the eigenvector

$$\mathbf{x} = \begin{bmatrix} 5 \\ -6 \end{bmatrix}.$$

Exercises 4.1

1. Let $A = \begin{bmatrix} 2 & -12 \\ 1 & -5 \end{bmatrix}$, $\mathbf{u}_1 = \begin{bmatrix} 4 \\ 1 \end{bmatrix}$, and $\mathbf{u}_2 = \begin{bmatrix} 3 \\ 1 \end{bmatrix}$.

    (a) Show that $\mathbf{u}_1$ is an eigenvector for A by calculating $A\mathbf{u}_1$. What is the eigenvalue, λ_1, corresponding to the eigenvector $\mathbf{u}_1$?

    (b) Graph the vectors $\mathbf{u}_1$ and $A\mathbf{u}_1$.

    (c) Show that $\mathbf{u}_2$ is an eigenvector for A by calculating $A\mathbf{u}_2$. What is the eigenvalue, λ_2, corresponding to the eigenvector $\mathbf{u}_2$?

    (d) Graph the vectors $\mathbf{u}_2$ and $A\mathbf{u}_2$.

2. Repeat Exercise 1 for $A = \begin{bmatrix} 1 & 2 \\ -1 & 4 \end{bmatrix}$, $\mathbf{u}_1 = \begin{bmatrix} 2 \\ 1 \end{bmatrix}$, and $\mathbf{u}_2 = \begin{bmatrix} 1 \\ 1 \end{bmatrix}$.

In Exercises 3-13 find the eigenvalues for the given matrix. For each eigenvalue, find a corresponding eigenvector.

3. $\begin{bmatrix} 5 & -1 \\ 2 & 2 \end{bmatrix}$ 4. $\begin{bmatrix} 2 & 2 \\ 2 & -1 \end{bmatrix}$ 5. $\begin{bmatrix} 3 & 0 \\ 5 & -2 \end{bmatrix}$

6. $\begin{bmatrix} 2 & -1 \\ 0 & 2 \end{bmatrix}$ 7. $\begin{bmatrix} 4 & 0 \\ 0 & 4 \end{bmatrix}$ 8. $\begin{bmatrix} 2 & 0 & 0 \\ 6 & -4 & -1 \\ -5 & 5 & 2 \end{bmatrix}$

9. $\begin{bmatrix} 4 & 0 & -2 \\ 5 & 3 & 0 \\ 6 & 0 & -4 \end{bmatrix}$ 10. $\begin{bmatrix} 4 & 2 & 4 \\ 0 & 3 & 1 \\ 0 & 0 & 2 \end{bmatrix}$ 11. $\begin{bmatrix} 3 & 1 & 2 \\ 0 & 3 & 1 \\ 0 & 0 & 3 \end{bmatrix}$

12. $\begin{bmatrix} 3 & 1 & 2 \\ 0 & 3 & 0 \\ 0 & 0 & 3 \end{bmatrix}$ 13. $\begin{bmatrix} 3 & 0 & 0 \\ 0 & 3 & 0 \\ 0 & 0 & 3 \end{bmatrix}$

14. For $A = \begin{bmatrix} 2 & x \\ 1 & -5 \end{bmatrix}$ and $\mathbf{u} = \begin{bmatrix} 1 \\ -1 \end{bmatrix}$, find x so that $\mathbf{u}$ is an eigenvector for A. What is the corresponding eigenvalue, λ?

15. For $A = \begin{bmatrix} x & y \\ 2x & -y \end{bmatrix}$ and $\mathbf{u} = \begin{bmatrix} 1 \\ -1 \end{bmatrix}$, find x and y so that $\mathbf{u}$ is an eigenvector corresponding to the eigenvalue $\lambda = 1$.

4.2 APPLICATIONS

This brief section describes an application where eigenvalues are used. The application involves a _difference equation_. Such equations are used in the study of population dynamics, ecological systems, digital control of chemical processes, radar tracking of airborne objects, and the like.

A Markov chain

Example 1 An automobile rental company has three locations which we designate as P, Q, and R. When an automobile is rented at one of the locations, it may be returned to any of the three locations.

At the start of a given week, suppose there are p cars at location P, q cars at location Q, and r cars at location R. By the end of the week, experience has shown that the p cars starting at location P will be distributed as follows:

> 10% are rented and returned to Q
> 30% are rented and returned to R
> 60% remain at P (either returned to P or not rented)

Similar rental histories are known for locations Q and R. All the histories are summarized below.

Weekly Distribution History

Location _P_: 60% stay at P	10% go to Q	30% go to R
Location _Q_: 10% go to P	80% stay at Q	10% go to R
Location _R_: 10% go to P	20% go to Q	70% stay at R

Suppose the automobile rental company has a fleet of 600 cars, and wants to start the year with 200 cars at each location. Estimate the number of cars at each location when the year ends.

Remarks: The situation described in Example 1 is an example of a _**Markov chain**_—a fixed population (the rental car fleet) is rearranged in stages (week by week) among a fixed number of categories (locations P, Q, and R). A Markov chain such as the one described in Example 1 can be modeled as an iteration of the form:

(1)
$$\mathbf{x}_{k+1} = A\mathbf{x}_k \, , k = 0, 1, 2, \ldots$$

In Equation (1), the vector $\mathbf{x}_i$ gives the _**state**_ of the system at the ith stage; $\mathbf{x}_i$ is called a _**state vector**_. In Example 1, the state vector has the form

$$\mathbf{x}_i = \begin{bmatrix} p(i) \\ q(i) \\ r(i) \end{bmatrix}$$

183

where:

> $p(i)$ denotes the number of cars at location P at the start of week i
> $q(i)$ denotes the number of cars at location Q at the start of week i
> $r(i)$ denotes the number of cars at location R at the start of week i

The matrix A in Equation (1) is called the **_state transition matrix_**. As can be seen from Equation (1), the matrix A allows us to calculate x_{k+1} (the state of the system at time $k + 1$) if we know x_k (the state of the system at time k).

The state equations for Example 1

We can use the weekly distribution history for Example 1 to derive the **_state equations_** for the rental car fleet. In particular, the components of the state vector are given by:

$$p(k + 1) = 0.6p(k) + 0.1q(k) + 0.1r(k)$$

$$q(k + 1) = 0.1p(k) + 0.8q(k) + 0.2r(k)$$

$$r(k + 1) = 0.3p(k) + 0.1q(k) + 0.7r(k).$$

Converting the state equations to vector form, we have

(2a)
$$x_{k+1} = Ax_k,\ k = 0, 1, 2, \ldots$$

where

(2b)
$$\mathbf{x}_k = \begin{bmatrix} p(k) \\ q(k) \\ r(k) \end{bmatrix}, \qquad A = \begin{bmatrix} 0.6 & 0.1 & 0.1 \\ 0.1 & 0.8 & 0.2 \\ 0.3 & 0.1 & 0.7 \end{bmatrix}.$$

Evolution of the state vectors in Example 1

Example 1 posed the question: "How many cars are at each location after one year?" In terms of the state equations, this question is answered by calculating the state vector x_{52}, where x_i and A are as in Equation (2) and where the initial state vector x_0 is given by

$$\mathbf{x}_0 = \begin{bmatrix} 200 \\ 200 \\ 200 \end{bmatrix}.$$

To find x_{52} we could proceed in stages, finding first x_1, then x_2:

$$x_1 = Ax_0 = \begin{bmatrix} 160 \\ 220 \\ 220 \end{bmatrix} \quad , \quad x_2 = Ax_1 = \begin{bmatrix} 140 \\ 236 \\ 224 \end{bmatrix} \quad , \text{ etc.}$$

These calculations show how the state of the system evolves. For instance, x_2 tells us there are 140 cars at location P, 236 at location Q, and 224 at location R at week 2. We are looking for x_{52}, however, and there is a way to calculate x_{52} without calculating all the intermediate state vectors $x_1, x_2, \ldots$.

The state vector x_i is given by $x_i = A^i x_0$

We now observe that the vectors x_i defined by Equation (1) can be expressed in the form

$$x_i = A^i x_0 \; , \; i = 0, 1, 2, \ldots$$

By way of justification, consider

$$x_1 = Ax_0$$

$$x_2 = Ax_1 = A(Ax_0) = A^2 x_0$$

$$x_3 = Ax_2 = A(A^2 x_0) = A^3 x_0 \; , \text{ etc.}$$

The solution of Example 1

To find the solution for Example 1, we will calculate x_{52} from $x_{52} = A^{52} x_0$. Using linear algebra software (such as MATLAB or Mathematica), we find

$$A^{52} = \begin{bmatrix} 0.2 & 0.2 & 0.2 \\ 0.45 & 0.45 & 0.45 \\ 0.35 & 0.35 & 0.35 \end{bmatrix}.$$

Having A^{52} we can now find x_{52}:

$$x_{52} = A^{52} x_0 = \begin{bmatrix} 0.2 & 0.2 & 0.2 \\ 0.45 & 0.45 & 0.45 \\ 0.35 & 0.35 & 0.35 \end{bmatrix} \begin{bmatrix} 200 \\ 200 \\ 200 \end{bmatrix} = \begin{bmatrix} 120 \\ 270 \\ 210 \end{bmatrix}.$$

Thus, to the extent that Equation (2) is valid, we expect to see more cars at location Q by the end of the year. Therefore, in planning for facilities, we need to allow for more parking and for a larger maintenance area at location Q.

In terms of planning, we now have to ask ourselves if the trend exhibited above will continue. That is, as time passes will there always be more cars at location Q? Or might the trend reverse at a later time and might we see more cars at location P or at location R? Put mathematically, we want to compute the following limit

(3)
$$\lim_{k \to \infty} \mathbf{x}_k.$$

If the limit (3) exists, the limit vector is called the _steady state_. In the next subsection we will see how to use eigenvalues and eigenvectors to find the steady state. Our analysis will show, relative to the rental car company in Example 1, that the fleet population does not oscillate but rather tends to a steady state of 120 cars at P, 270 cars at Q, and 210 cars at R. This knowledge is of obvious value in terms of facilities planning.

Calculating the limit of $\mathbf{x}_k$ as k tends to infinity

We turn to the general iteration given in Equation (1):

$$\mathbf{x}_{k+1} = A\mathbf{x}_k, \, k = 0, 1, 2, \ldots$$

As we pointed out earlier, we can calculate the ith vector, $\mathbf{x}_i$, from

$$\mathbf{x}_i = A^i\mathbf{x}_0.$$

Thus, computing the limit of $\mathbf{x}_i$ as i tends to infinity amounts to computing the limit of $A^i\mathbf{x}_0$ as i tends to infinity.

Now, while it is hard to calculate $A^i\mathbf{x}$ for an arbitrary vector $\mathbf{x}$, it turns out to be very easy to calculate $A^i\mathbf{x}$ if $\mathbf{x}$ is an eigenvector of A. For example, suppose $A\mathbf{x} = \lambda\mathbf{x}$. Then

$$A^2\mathbf{x} = A(A\mathbf{x}) = A(\lambda\mathbf{x}) = \lambda(A\mathbf{x}) = \lambda(\lambda\mathbf{x}) = \lambda^2\mathbf{x}$$

$$A^3\mathbf{x} = A(A^2\mathbf{x}) = A(\lambda^2\mathbf{x}) = \lambda^2(A\mathbf{x}) = \lambda^2(\lambda\mathbf{x}) = \lambda^3\mathbf{x},$$

and, in general,

(4)
$$A^i\mathbf{x} = \lambda^i\mathbf{x}, \, i = 1, 2, 3, \ldots$$

So, rather than trying to calculate $A^i\mathbf{x}_0$ for an arbitrary starting vector $\mathbf{x}_0$, we instead represent $\mathbf{x}_0$ in terms of eigenvectors and use Equation (4).

To begin, we assume A is an $(n \times n)$ matrix and that A has a set of n eigenvalues and eigenvectors where:

$$Au_1 = \lambda_1 u_1$$

$$Au_2 = \lambda_2 u_2$$

$$Au_n = \lambda_n u_n \ .$$

Next, let us suppose that we can represent the starting vector x_0 in terms of u_1, $u_2, \ldots, u_n$ as follows:

(5) $$x_0 = c_1 u_1 + c_2 u_2 + \cdots + c_n u_n.$$

Then,

$$x_i = A^i x_0$$

(6)
$$= A^i (c_1 u_1 + c_2 u_2 + \cdots + c_n u_n)$$

$$= c_1 A^i u_1 + c_2 A^i u_2 + \cdots + c_n A^i u_n$$

$$= c_1 \lambda_1^i u_1 + c_2 \lambda_2^i u_2 + \cdots + c_n \lambda_n^i u_n \ .$$

The equation above tells us that the limit of x_i depends on limits of powers of the different eigenvalues. To provide some structure for the analysis, let us assume that the eigenvalues are ordered as follows:

$$|\lambda_1| \ge |\lambda_2| \ge \cdots \ge |\lambda_n|.$$

Then, from (6), some of the possibilities are:

1. If $|\lambda_1| < 1$, then $\lim_{i \to \infty} x_i = 0$

2. If $\lambda_1 = 1$ and $|\lambda_2| < 1$, then $\lim_{i \to \infty} x_i = c_1 u_1$

3. If $|\lambda_1| > 1$, then $\lim_{i \to \infty} \|x_i\| = \infty$

There are possibilities besides the ones listed above. For example, we don't have the case $\lambda_1 = -1$, or the case $\lambda_1 = \lambda_2 = 1$, etc. None the less, it should be clear how to use (6) to find the limit of x_i as i tends to infinity.

A complete analysis for Example 1

The state vectors for the rental fleet of Example 1 are determined by Equation (2). The state transition matrix A has eigenvalues and eigenvectors given by:

$$\lambda_1 = 1 \, , \, \mathbf{u}_1 = \begin{bmatrix} 4 \\ 9 \\ 7 \end{bmatrix} \; ; \quad \lambda_2 = 0.6 \, , \, \mathbf{u}_2 = \begin{bmatrix} 0 \\ 1 \\ -1 \end{bmatrix} \; ; \quad \lambda_3 = 0.5 \, , \, \mathbf{u}_3 = \begin{bmatrix} -1 \\ -1 \\ 2 \end{bmatrix}.$$

The initial state vector $\mathbf{x}_0$ can be represented in terms of the eigenvectors as follows:

$$\mathbf{x}_0 = \begin{bmatrix} 200 \\ 200 \\ 200 \end{bmatrix} = 30 \begin{bmatrix} 4 \\ 9 \\ 7 \end{bmatrix} - 150 \begin{bmatrix} 0 \\ 1 \\ -1 \end{bmatrix} - 80 \begin{bmatrix} -1 \\ -1 \\ 2 \end{bmatrix}.$$

Therefore,

$$\mathbf{x}_i = A^i \mathbf{x}_0 = 30(1)^i \begin{bmatrix} 4 \\ 9 \\ 7 \end{bmatrix} - 150(0.6)^i \begin{bmatrix} 0 \\ 1 \\ -1 \end{bmatrix} - 80(0.5)^i \begin{bmatrix} -1 \\ -1 \\ 2 \end{bmatrix}$$

$$= \begin{bmatrix} 120 \\ 270 \\ 210 \end{bmatrix} - 150(0.6)^i \begin{bmatrix} 0 \\ 1 \\ -1 \end{bmatrix} - 80(0.5)^i \begin{bmatrix} -1 \\ -1 \\ 2 \end{bmatrix}.$$

From above, it is clear that the fleet will tend to a distribution of 120 cars at P, 270 cars at Q, and 210 cars at R. In fact, no matter what the initial size of the fleet is and no matter how they are based at week 0, the final distribution will be in the proportion of 4 to 9 to 7, the same proportion as the components of the dominant eigenvector $\mathbf{u}_1$. That is, if the initial state vector $\mathbf{x}_0$ is given by

(7) $$\mathbf{x}_0 = c_1 \mathbf{u}_1 + c_2 \mathbf{u}_2 + c_3 \mathbf{u}_3,$$

then

$$\mathbf{x}_i = c_1 \mathbf{u}_1 + c_2 (0.6)^i \mathbf{u}_2 + c_3 (0.5)^i \mathbf{u}_3.$$

Therefore, as i grows larger, the state vector $\mathbf{x}_i$ lines up more and more along the steady state $c_1 \mathbf{u}_1$. (Suppose that $\mathbf{x}_0$ is a "feasible" state vector; one with no negative components and with at least one positive component. Then, it is interesting to note, we can prove that the coefficient c_1 is positive. Put another way, if c_1 is zero or negative, then we can show that either $\mathbf{x}_0$ has some negative components or $\mathbf{x}_0 = \mathbf{0}$.)

Exercises 4.2

1. The matrix $A = \begin{bmatrix} 2 & 1/2 \\ -3 & -1/2 \end{bmatrix}$ has eigenvalues $\lambda_1 = 1$ and $\lambda_2 = 1/2$ with corresponding eigenvectors $\mathbf{u_1} = \begin{bmatrix} 1 \\ -2 \end{bmatrix}$ and $\mathbf{u_2} = \begin{bmatrix} 1 \\ -3 \end{bmatrix}$, resectively. Set $\mathbf{x_0} = \begin{bmatrix} 1 \\ 0 \end{bmatrix}$ and let $\mathbf{x_i} = A^i \mathbf{x_0}$ for $i = 1, 2, 3 \ldots$.

   (a) Find scalars a_1 and a_2 so that $\mathbf{x_0} = a_1 \mathbf{u_1} + a_2 \mathbf{u_2}$.

   (b) Find scalars b_1 and b_2 so that $\mathbf{x_1} = b_1 \mathbf{u_1} + b_2 \mathbf{u_2}$.

   (c) Find scalars c_1 and c_2 so that $\mathbf{x_5} = c_1 \mathbf{u_1} + c_2 \mathbf{u_2}$.

   (d) Determine $\lim_{k \to \infty} \mathbf{x_k}$.

2. Two newspapers compete for subscriptions in a region with 300,000 households. Assume that no household subscribes to both newspapers and that the table below gives the probabilities that a household will change its subscription status during the year.

	From A	From B	From None
To A	.70	.15	.30
To B	.20	.80	.20
To None	.10	.05	.50

   For example, an interpretation of the first column of the table is that during a given year, newspaper A can expect to keep 70% of its current subscribers while losing 20% to newspaper B and 10% to no subscription.

   At the beginning of a particular year, suppose that 150,000 households subscribe to newspaper A, 100,000 subscribe to newspaper B, and 50,000 have no subscription. Let P and $\mathbf{x_0}$ be defined by

$$P = \begin{bmatrix} .70 & .15 & .30 \\ .20 & .80 & .20 \\ .10 & .05 & .50 \end{bmatrix} \quad \text{and} \quad \mathbf{x_0} = \begin{bmatrix} 150,000 \\ 100,000 \\ 50,000 \end{bmatrix}.$$

189

For $i = 1, 2, \ldots$, set $\mathbf{x}_i = P^i\mathbf{x}_0$

(a) Calculate $\mathbf{x}_1$ and $\mathbf{x}_2$ and interpret the results.

(b) Verify that $\mathbf{x}_0 = 1.9121\mathbf{u}_1 + 0.6374\mathbf{u}_2$, where
$$\mathbf{u}_1 = \begin{bmatrix} 58835 \\ 78446 \\ 19612 \end{bmatrix} \text{ and } \mathbf{u}_2 = \begin{bmatrix} 58835 \\ -78446 \\ 19612 \end{bmatrix}.$$

(c) Given that $\mathbf{u}_1$ and $\mathbf{u}_2$ are eigenvectors for P corresponding to the eigenvalues $\lambda_1 = 1$ and $\lambda_2 = 0.6$, respectively, determine $\lim_{k \to \infty} \mathbf{x_k}$.

4.3 EIGENVALUES OF SPECIAL MATRICES

In this section we state a number of important results relative to solutions of the eigenvalue problem. We begin by observing that a matrix might have complex eigenvalues.

Some matrices have complex eigenvalues

For a given matrix A, the eigenvalues of A are the roots of the characteristic equation, $p(\lambda) = 0$. However, as we know from elementary algebra, a polynomial equation sometimes has complex roots.

Therefore, we might expect that certain matrices can have complex eigenvalues. The following example shows that complex eigenvalues are indeed a possibility.

Example 1   Find the eigenvalues for $A = \begin{bmatrix} 3 & 1 \\ -2 & 1 \end{bmatrix}$.

Solution: Calculating the characteristic polynomial, we obtain

$$p(\lambda) = \begin{vmatrix} 3-\lambda & 1 \\ -2 & 1-\lambda \end{vmatrix}$$

$$= \lambda^2 - 4\lambda + 5$$

We apply the quadratic formula to the characteristic equation $\lambda^2 - 4\lambda + 5 = 0$, finding eigenvalues:

$$\lambda = \frac{4 \pm \sqrt{16-20}}{2}$$

$$= \frac{4 \pm \sqrt{-4}}{2}$$

$$= \frac{4 \pm 2i}{2}$$

$$= 2 \pm i.$$

Thus, the eigenvalues of A are the complex numbers $\lambda = 2 + i$ and $\lambda = 2 - i$, where i is the imaginary number $i = \sqrt{-1}$. ∎

Eigenvectors corresponding to complex eigenvalues are also complex

If a real matrix A has complex eigenvalues (as in Example 1), you might suspect that the corresponding eigenvectors are complex as well. Indeed, this suspicion is correct.

In Section 4 we review complex numbers and their arithmetic as a prelude to studying the complications caused by complex eigenvalues.

A symmetric matrix never has a complex eigenvalue

The following theorem shows us, for certain matrices, that we never have to worry about the possibility of complex eigenvalues.

THEOREM 1　　　　　Let A be a symmetric $(n \times n)$ matrix. Then every eigenvalue of A is real.

Proof: We give an elementary proof of Theorem 1 for a (2×2) symmetric matrix A. In Section 5, we establish Theorem 1 for an $(n \times n)$ matrix.

If A is a (2×2) symmetric matrix, then A has the form

$$A = \begin{bmatrix} p & q \\ q & r \end{bmatrix}.$$

Therefore, the characteristic polynomial for A is

$$p(\lambda) = \lambda^2 - (p + r)\lambda + pr - q^2.$$

Applying the quadratic formula to find the eigenvalues, we obtain

$$\lambda = \frac{(p+r) \pm \sqrt{(p+r)^2 - 4(pr - q^2)}}{2}$$

$$= \frac{(p+r) \pm \sqrt{(p-r)^2 + 4q^2}}{2}$$

Since the quantity under the radical is never negative, we see that the eigenvalue can never be complex.　∎

Before turning to Section 4 and an introduction to complex numbers, we list several other results about the eigenvalue problem.

The eigenvalues of a triangular matrix are its main diagonal entries

It can be shown that the determinant of a triangular matrix is the product of its main diagonal entries. In adddition, note that if A is a triangular matrix, then so is $A - \lambda I$.

Combining these two observations, it follows that the eigenvalues of an $(n \times n)$ triangular matrix A are the main diagonal entries $a_{11}, a_{22}, \ldots, a_{nn}$. We prove this fact for the special case of a (3×3) lower-triangular matrix. In particular,

$$p(\lambda) = \det(A - \lambda I)$$

$$= \begin{vmatrix} a_{11} - \lambda & 0 & 0 \\ a_{21} & a_{22} - \lambda & 0 \\ a_{31} & a_{32} & a_{33} - \lambda \end{vmatrix}$$

$$= (a_{11} - \lambda)(a_{22} - \lambda)(a_{33} - \lambda)$$

This calculation shows that the eigenvalues of a (3×3) lower-triangular matrix are its diagonal entries. A similar proof will work for any $(n \times n)$ lower-triangular matrix or any $(n \times n)$ upper-triangular matrix.

The number $\lambda = 0$ is an eigenvalue of A if and only if A is not invertible

Although the vector $\mathbf{x} = \mathbf{0}$ is never an eigenvector, the number $\lambda = 0$ can be an eigenvalue. In particular, let $p(\lambda) = \det(A - \lambda I)$ be the characteristic polynomial for a square matrix A. Observe that $p(0) = 0$ if and only if $\det(A) = 0$. Therefore (since invertible matrices always have nonzero determinants), $\lambda = 0$ is an eigenvalue of A if and only if A is not invertible.

__Example 2__   Find the eigenvalues and eigenvectors for $A = \begin{bmatrix} 2 & 3 \\ 4 & 6 \end{bmatrix}$.

__Solution:__   The characteristic polynomial is

$$p(\lambda) = \begin{vmatrix} 2 - \lambda & 3 \\ 4 & 6 - \lambda \end{vmatrix}$$

$$= \lambda^2 - 8\lambda$$

$$= \lambda(\lambda - 8).$$

193

Thus, the eigenvalues of A are $\lambda = 0$ and $\lambda = 8$. (Note that $\det(A) = 0$ and hence A is not invertible.)

The eigenvectors corresponding to $\lambda = 0$ and $\lambda = 8$ are, respectively:

$$\mathbf{x} = \begin{bmatrix} x_1 \\ x_2 \end{bmatrix} = x_2 \begin{bmatrix} -3/2 \\ 1 \end{bmatrix} \quad , \quad \mathbf{x} = \begin{bmatrix} x_1 \\ x_2 \end{bmatrix} = x_2 \begin{bmatrix} 1/2 \\ 1 \end{bmatrix} \quad , \quad x_2 \neq 0.$$

■

A and A^T have the same eigenvalues

We know from Section 2.9 that a matrix and its transpose have the same determinant. As we see from the sequence of equalities below, a matrix and its transpose also have the same characteristic polynomial:

$$\det(A - \lambda I) = \det((A - \lambda I)^T) = \det(A^T - \lambda I^T) = \det(A^T - \lambda I).$$

As is shown above, A and A^T have the same characteristic polynomial. Therefore, A and A^T have precisely the same eigenvalues. As is illustrated in Example 3, however, we cannot necessarily expect that A and A^T have the same eigenvectors.

Example 3   Find the eigenvalues and eigenvectors for A and A^T:

$$A = \begin{bmatrix} 2 & 4 & 4 \\ 0 & 1 & -1 \\ 0 & 1 & 3 \end{bmatrix} \quad , \quad A^T = \begin{bmatrix} 2 & 0 & 0 \\ 4 & 1 & 1 \\ 4 & -1 & 3 \end{bmatrix}.$$

Solution: We only calculate the characteristic polynomial for A and leave it for you to show that A^T has exactly the same characteristic polynomial. (We calculate $p(\lambda)$ using a first column expansion for the determinant.)

$$p(\lambda) = \begin{vmatrix} 2 - \lambda & 4 & 4 \\ 0 & 1 - \lambda & -1 \\ 0 & 1 & 3 - \lambda \end{vmatrix}$$

$$= (2 - \lambda) \begin{vmatrix} 1 - \lambda & -1 \\ 1 & 3 - \lambda \end{vmatrix}$$

$$= (2 - \lambda)(\lambda^2 - 4\lambda + 4)$$

$$= -(\lambda - 2)^3$$

194

Thus, the only eigenvalue for A is $\lambda = 2$.

To find the corresponding eigenvectors, we form the augmented matrix for $(A - 2I)\mathbf{x} = \mathbf{0}$ and transform it to reduced echelon form:

$$[A - 2I \mid \mathbf{0}] = \begin{bmatrix} 0 & 4 & 4 & 0 \\ 0 & -1 & -1 & 0 \\ 0 & 1 & 1 & 0 \end{bmatrix} \longrightarrow \begin{bmatrix} 0 & 1 & 1 & 0 \\ 0 & 0 & 0 & 0 \\ 0 & 0 & 0 & 0 \end{bmatrix}.$$

Therefore, the system $(A - 2I)\mathbf{x} = \mathbf{0}$ reduces to the equivalent system

$$x_2 + x_3 = 0,$$

or

$$x_2 = -x_3.$$

Consequently, eigenvectors corresponding to $\lambda = 2$ are nonzero vectors of the form

$$\mathbf{x} = \begin{bmatrix} x_1 \\ x_2 \\ x_3 \end{bmatrix} = \begin{bmatrix} x_1 \\ -x_3 \\ x_3 \end{bmatrix} = x_1 \begin{bmatrix} 1 \\ 0 \\ 0 \end{bmatrix} + x_3 \begin{bmatrix} 0 \\ -1 \\ 1 \end{bmatrix}.$$

We have seen that A and A^T have exactly the same characteristic polynomial, $p(\lambda) = -(\lambda - 2)^3$. To find the eigenvectors for A^T corresponding to the eigenvalue $\lambda = 2$ we solve $(A^T - 2I)\mathbf{x} = \mathbf{0}$. Forming the augmented matrix and reducing it, we find that this system is equivalent to

$$x_1 = (x_2 - x_3)/4.$$

Therefore, the corresponding eigenvectors are nonzero vectors of the form:

$$\mathbf{x} = \begin{bmatrix} x_1 \\ x_2 \\ x_3 \end{bmatrix} = \begin{bmatrix} \frac{1}{4}(x_2 - x_3) \\ x_2 \\ x_3 \end{bmatrix} = x_2 \begin{bmatrix} 1/4 \\ 1 \\ 0 \end{bmatrix} + x_3 \begin{bmatrix} -1/4 \\ 0 \\ 1 \end{bmatrix}.$$

So, even though A and A^T always have the same eigenvalues, this example illustrates that they will likely have different eigenvectors. ∎

Exercises 4.3

In Exercises 1-4, find the eigenvalues for the given matrix.

1. $\begin{bmatrix} 1 & 2 \\ -2 & 1 \end{bmatrix}$

2. $\begin{bmatrix} 0 & -1 \\ 1 & 0 \end{bmatrix}$

3. $\begin{bmatrix} 2 & 0 & 0 \\ 1 & 3 & 5 \\ 2 & -1 & -1 \end{bmatrix}$

4. $\begin{bmatrix} 0 & 0 & 0 \\ 0 & 0 & -1 \\ 0 & 1 & 0 \end{bmatrix}$

5. Show that for real numbers a and b, with $b \neq 0$, the matrix $A = \begin{bmatrix} a & b \\ -b & a \end{bmatrix}$ has eigenvalues $a \pm bi$ with corresponding eigenvectors $\begin{bmatrix} 1 \\ i \end{bmatrix}$, and $\begin{bmatrix} 1 \\ -i \end{bmatrix}$.

6. Show that for real numbers a and b, the matrix $A = \begin{bmatrix} a & b \\ b & a \end{bmatrix}$ has eigenvalues $a \pm b$ with corresponding eigenvectors $\begin{bmatrix} 1 \\ 1 \end{bmatrix}$, and $\begin{bmatrix} 1 \\ -1 \end{bmatrix}$.

7. Find the eigenvalues and corresponding eigenvectors for the matrix $A = \begin{bmatrix} 3 & 0 & 0 \\ 1 & 2 & 0 \\ -1 & 2 & -1 \end{bmatrix}$.

8. Find the eigenvalues and corresponding eigenvectors for the matrix $A = \begin{bmatrix} 3 & -2 & 1 \\ 0 & 3 & -2 \\ 0 & 0 & 2 \end{bmatrix}$.

9. Let $A = \begin{bmatrix} 3 & 2 \\ 3 & -2 \end{bmatrix}$.

    (a) Find the eigenvalues and corresponding eigenvectors for A.

(b) Find the eigenvalues and corresponding eigenvectors for A^T.

10. The matrix $A = \begin{bmatrix} 4 & 2 \\ -1 & 1 \end{bmatrix}$ has eigenvalues $\lambda_1 = 2$ and $\lambda_2 = 3$ with corresponding eigenvectors $\mathbf{u}_1 = \begin{bmatrix} 1 \\ -1 \end{bmatrix}$ and $\mathbf{u}_2 = \begin{bmatrix} 2 \\ -1 \end{bmatrix}$, respectively.

(a) Find A^{-1} and verify that A^{-1} has eigenvalues $1/\lambda_1$ and $1/\lambda_2$ with corresponding eigenvectors $\mathbf{u}_1 = \begin{bmatrix} 1 \\ -1 \end{bmatrix}$ and $\mathbf{u}_2 = \begin{bmatrix} 2 \\ -1 \end{bmatrix}$, respectively.

(b) Find $2A$ and verify that $2A$ has eigenvalues $2\lambda_1$ and $2\lambda_2$ with corresponding eigenvectors $\mathbf{u}_1 = \begin{bmatrix} 1 \\ -1 \end{bmatrix}$ and $\mathbf{u}_2 = \begin{bmatrix} 2 \\ -1 \end{bmatrix}$, respectively.

(c) Find A^2 and verify that A^2 has eigenvalues λ_1^2 and λ_2^2 with corresponding eigenvectors $\mathbf{u}_1 = \begin{bmatrix} 1 \\ -1 \end{bmatrix}$ and $\mathbf{u}_2 = \begin{bmatrix} 2 \\ -1 \end{bmatrix}$, respectively.

11. Let A be an $(n \times n)$ matrix and suppose λ is an eigenvalue for A with associated eigenvector $\mathbf{u}$.

(a) If A is invertible (hence $\lambda \neq 0$), show that $1/\lambda$ is an eigenvalue for A^{-1} with associated eigenvector $\mathbf{u}$.

(b) If k is a scalar, show that $k\lambda$ is an eigenvalue for kA with associated eigenvector $\mathbf{u}$.

(a) Show that λ^2 is an eigenvalue for A^2 with associated eigenvector $\mathbf{u}$.

197

4.4 COMPLEX NUMBERS AND COMPLEX ARITHMETIC

As we saw in the previous section, a given matrix might have complex eigenvalues. Therefore, we need to know a little complex arithmetic so that we can find the corresponding eigenvectors.

Complex numbers

A number such as

$$z = 3 + 5i$$

is called a *complex number*. The symbol i stands for $\sqrt{-1}$. That is, i is a number having the property that $i^2 = -1$.

In general, let a and b be real numbers. A number of the form

$$z = a + ib$$

is called a ***complex number***. In the representation $z = a + ib$, the number a is called the ***real part*** of z while the number b is the ***imaginary part*** of z.

For example, let $z = 3 + 5i$. The real part of z is 3 and the imaginary part of z is 5.

The complex plane

A complex number has two "components," its real part and its imaginary part. Therefore, it is customary to represent a complex number $z = a + ib$ as a point in the ***complex plane*** having coordinates (a, b), see Figure 1.

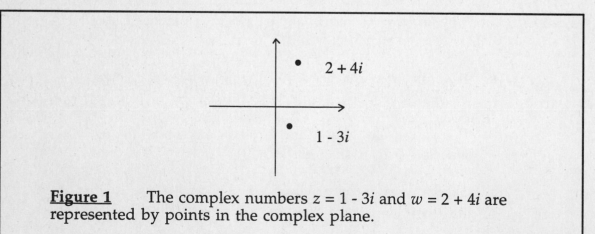

Figure 1   The complex numbers $z = 1 - 3i$ and $w = 2 + 4i$ are represented by points in the complex plane.

Adding and multiplying complex numbers

We add complex numbers by individually adding their real parts and their imaginary parts. In particular, let $z_1 = a_1 + ib_1$ and $z_2 = a_2 + ib_2$ be complex numbers. The **_sum_**, $z_1 + z_2$, is defined as follows:

$$z_1 + z_2 = (a_1 + ib_1) + (a_2 + ib_2) = (a_1 + a_2) + i(b_1 + b_2).$$

We multiply z_1 and z_2 by forming the product

$$z_1 z_2 = (a_1 + ib_1)(a_2 + ib_2) = a_1(a_2 + ib_2) + ib_1(a_2 + ib_2)$$

$$= a_1 a_2 + ia_1 b_2 + ia_2 b_1 + i^2 b_1 b_2.$$

Using the fact that $i^2 = -1$ and collecting terms involving i, we are led to define the **_product_**, $z_1 z_2$, as follows:

$$z_1 z_2 = (a_1 + ib_1)(a_2 + ib_2) = (a_1 a_2 - b_1 b_2) + i(a_1 b_2 + a_2 b_1).$$

The example below illustrates complex addition and multiplication.

Example 1   Let $z = 2 + 3i$ and let $w = 2 - 5i$. Form the sum $z + w$ and the product zw.

Solution: We find

$$z + w = (2 + 3i) + (2 - 5i) = 4 - 2i$$

and

$$zw = (2 + 3i)(2 - 5i) = 19 - 4i.$$

■

Dividing complex numbers

The division process for complex numbers is not quite so straightforward as addition and multiplication. At first glance, for example, the following calculation of 2 divided by $1 + i$ may look a bit strange:

$$\frac{2}{1+i} = 1 - i.$$

(To verify the division calculation displayed above, cross multiply and check that $2 = (1 + i)(1 - i)$.)

Dividing complex numbers is similar to the high school algebra process for rationalizing fractions that contain square roots. In particular, consider complex numbers $z_1 = a_1 + ib_1$ and $z_2 = a_2 + ib_2$. We form the **_quotient_**, z_1 / z_2, as follows:

$$\frac{z}{z_2} = \frac{a_1 + ib_1}{a_2 + ib_2} = \frac{a_1 + ib_1}{a_2 + ib_2} \frac{a_2 - ib_2}{a_2 - ib_2}$$

$$= \frac{(a_1 + ib_1)(a_2 - ib_2)}{(a_2)^2 + (b_2)^2}$$

The point of the calculation above is to get the quotient, z_1/z_2, into the form of a complex number; that is, we need to express z_1/z_2 in the standard form $z_1/z_2 = c + id$. Since $(a_2)^2 + (b_2)^2$ is a real number, the "rationalization" process displayed above will get us to our objective.

Example 2:   Let $z = 3 + 5i$ and let $w = 3 - i$. Form the quotient z/w.

Solution: We find

$$\frac{z}{w} = \frac{3+5i}{3-i} = \frac{3+5i}{3-i} \frac{3+i}{3+i} = \frac{4+18i}{10} = 0.4 + 1.8i$$

■

The complex conjugate

The *conjugate* operation is quite useful when dealing with complex numbers. Let $z = a + ib$ denote a complex number. The **_complex conjugate_** of z, denoted by $\bar{z}$, is defined by

$$\overline{a+ib} = a - ib \ .$$

In other words, we form $\bar{z}$ by changing the sign of the imaginary part of z. For instance, if $z = 7 + 5i$, then $\bar{z} = 7 - 5i$.

In the complex plane, a complex number and its conjugate form pairs that are symmetric with respect to the horizontal axis, see Figure 2.

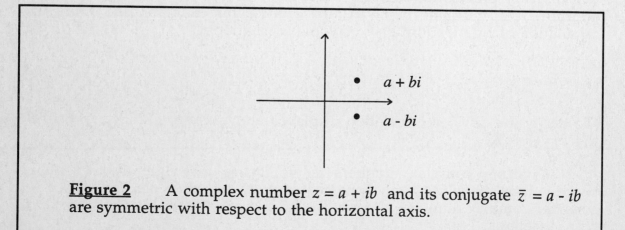

Figure 2   A complex number $z = a + ib$ and its conjugate $\bar{z} = a - ib$ are symmetric with respect to the horizontal axis.

Algebraic properties of the conjugate

The real numbers are a subset of the complex numbers. In particular, a complex number $z = a + ib$ is real if and only if the imaginary part, b, is zero. Another way to state this result is:

$$A \text{ complex number } z \text{ is real} \Leftrightarrow z = \bar{z} .$$

The test above is a valuable theoretical tool and we will use it later to show that the eigenvalues of a real symmetric matrix are always real.

The following theorem summarizes how the conjugate operation behaves when it is mixed in with addition and multiplication.

THEOREM 1   Let $z = a + ib$ and $w = c + id$ be complex numbers.

(a) $\overline{(z+w)} = \bar{z} + \bar{w}$ (b) $\overline{(zw)} = \bar{z}\,\bar{w}$

(c) $z + \bar{z} = 2a$ (d) $z\bar{z} = a^2 + b^2$

The proof of Theorem 1 is left to the exercises.

Finally, we note that the conjugate can be used to give an easily-remembered formula for complex division. In particular,

$$\frac{z_1}{z_2} = \frac{z_1 \bar{z}_2}{z_2 \bar{z}_2} .$$

The magnitude of a complex number

For a real number c, the absolute value of c measures how far c is from the origin of the real number line. Similarly, see Figure 2, if $z = a + ib$, then the distance from z to the origin of the complex plane is $\sqrt{a^2 + b^2}$. In particular, we define the *magnitude* of z, denoted $|z|$, by

$$|z| = \sqrt{a^2 + b^2} .$$

Using Part (d) of Theorem 1, we can also represent the magnitude of z as follows:

$$|z| = \sqrt{z\bar{z}} .$$

Example 3   Let $z = 2 + i$ and $w = 1 - 3i$. Find the magnitude of u, where u is given by $u = (z\overline{w})/(\overline{z} + w)$.

Solution: Calculating u we find

$$u = \frac{(2+i)(1+3i)}{(2-i)+(1-3i)} = \frac{-1+7i}{3-4i} = \frac{-1+7i}{3-4i}\frac{3+4i}{3+4i} = \frac{-31+17i}{25}.$$

Therefore,

$$|u| = \sqrt{(-31/25)^2 + (17/25)^2} = \sqrt{1250/625} = \sqrt{2}.$$

■

Exercises 4.4

In the Exercises for this section $u = 3 - 2i$, $v = 4 + i$, $w = 2 - i$.

1. Find the conjugates $\bar{u}$, $\bar{v}$, and $\bar{w}$.

2. Find $u\bar{u}$, $v\bar{v}$, and $w\bar{w}$.

3. Find $|u|$, $|v|$, and $|w|$.

In Exercises 4-18, express the given complex number in the form $a + ib$, where a and b are real numbers.

4. $u + v$ 5. $u + w$ 6. $u + \bar{u}$ 7. $u - \bar{u}$

8. $v + \bar{v}$ 9. $v - \bar{v}$ 10. uv 11. $u\bar{w}$

12. $\overline{\bar{u}v}$ 13. $v^2 + w$ 14. $u + iv$ 15. $v - iw$

16. u/v 17. u/w 18. $1/u$

19. Solve the system of equations

$$
\begin{array}{rcrcr}
x_1 & + & ix_2 & = & 1 \\
ix_1 & + & 2x_2 & = & -1
\end{array} .
$$

20. Solve the system of equations

$$
\begin{array}{rcrcr}
x_1 & + & ix_2 & = & 0 \\
(1 - i)x_1 & + & (1 + i)x_2 & = & 0
\end{array} .
$$

4.5 COMPLEX EIGENVECTORS

If A has an eigenvalue λ, then we find the corresponding eigenvectors by solving $(A - \lambda I)\mathbf{x} = \mathbf{0}$. In this section we examine the computational details of solving the eigenvector equation when λ is complex.

As we will see, Gauss-Jordan elimination works to solve complex systems in exactly the same way that works to solve real systems. The only significant difference is that the necessary complex arithmetic is more tiresome than real arithmetic.

An example of solving a complex system

Rather than giving an elaborate step-by-step description of how to solve a complex system, we will simply illustrate the process of Gauss-Jordan elimination with an example.

Observe that the elimination used in Example 1 proceeds exactly the same for the complex system as it would have for a real system—the only difference is that we use complex numbers as multipliers rather than real numbers.

Example 1   Find the eigenvalues and eigenvectors for $A = \begin{bmatrix} 3 & 1 \\ -2 & 1 \end{bmatrix}$.

Solution: We already know from Example 1 in Section 2 that the eigenvalues of A are $\lambda = 2 + i$ and $\lambda = 2 - i$.

We find the eigenvectors corresponding to $\lambda = 2 + i$ by solving the system $(A - (2 + i)I)\mathbf{x} = \mathbf{0}$. The augmented matrix for this system is

$$\begin{bmatrix} 1-i & 1 & 0 \\ -2 & -1-i & 0 \end{bmatrix}.$$

The first step in Gauss-Jordan elimination is to get a 1 in the $(1, 1)$ position. We accomplish this by multiplying row 1 by $1/(1 - i)$ where

$$\frac{1}{1-i} = \frac{1}{1-i}\frac{1+i}{1+i} = \frac{1+i}{2} = 0.5 + 0.5i.$$

Thus:

$$\begin{bmatrix} 1-i & 1 & 0 \\ -2 & -1-i & 0 \end{bmatrix} \xrightarrow{(0.5+0.5i)R_1} \begin{bmatrix} 1 & 0.5+0.5i & 0 \\ -2 & -1-i & 0 \end{bmatrix}$$

The next step is to introduce a 0 in the (2, 1) position:

$$\begin{bmatrix} 1 & 0.5+0.5i & 0 \\ -2 & -1-i & 0 \end{bmatrix} \xrightarrow{R_2+2R_1} \begin{bmatrix} 1 & 0.5+0.5i & 0 \\ 0 & 0 & 0 \end{bmatrix}.$$

Thus, the original complex system $(A - (2 + i)I)\mathbf{x} = \mathbf{0}$ is equivalent to the single equation:

$$x_1 + (0.5 + 0.5i)x_2 = 0,$$

or

$$x_1 = -(0.5 + 0.5i)x_2.$$

Therefore, the eigenvectors are given by

$$\mathbf{x} = \begin{bmatrix} x_1 \\ x_2 \end{bmatrix} = \begin{bmatrix} -(0.5+0.5i)x_2 \\ x_2 \end{bmatrix} = x_2 \begin{bmatrix} -0.5-0.5i \\ 1 \end{bmatrix}, \quad x_2 \neq 0.$$

A similar calculation shows the eigenvectors corresponding to $\lambda = 2 - i$ are given by

$$\mathbf{x} = x_2 \begin{bmatrix} -0.5+0.5i \\ 1 \end{bmatrix}, \quad x_2 \neq 0.$$

■

When eigenvectors are found by linear algebra software

We know that any nonzero multiple of an eigenvector is also an eigenvector. This means that we can eliminate fractions in an eigenvector by multiplying it by a constant.

For instance, in Example 1 we found the eigenvectors corresponding to $\lambda = 2 - i$ had the form

(1)
$$\mathbf{x} = x_2 \begin{bmatrix} -0.5-0.5i \\ 1 \end{bmatrix}, \quad x_2 \neq 0.$$

If we choose x_2 to be -2, the eigenvector $\mathbf{x}$ has the nicer form

$$\mathbf{x} = \begin{bmatrix} 1+i \\ -2 \end{bmatrix}.$$

Linear algebra software such as MATLAB or Mathematica will often choose an eigenvector that is normalized so that the first nonzero component is 1. We can do this for **x** in Equation (1) if we choose $x_2 = 1/(-0.5 - 0.5i)$. In particular:

$$\frac{1}{-0.5-0.5i} = \frac{1}{-0.5-0.5i}\frac{-0.5+0.5i}{-0.5+0.5i} = \frac{-0.5+0.5i}{0.5} = -1+i.$$

Choosing $x_2 = -1 + i$ in Equation (1) we find the eigenvector

$$\mathbf{x} = \begin{bmatrix} 1 \\ -1+i \end{bmatrix}.$$

The eigenvalues and eigenvectors of a real matrix occur in conjugate pairs

There are theoretical results that considerably simplify the eigenvalue problem. The most important fact is already familiar from Section 3. That is, if A is a symmetric matrix, then A cannot have a complex eigenvalue.

When complex eigenvalues are a possibility, however, Theorem 1 (which we state below) is useful. In words, it says that complex eigenvalues and eigenvectors occur in conjugate pairs—once we find one complex eigenvalue and eigenvector, then we get another free by taking the conjugate.

This fact is illustrated in Example 1 which found the following eigenvalue/eigenvector pairs:

$$\lambda = 2+i \quad , \quad \mathbf{x} = x_2 \begin{bmatrix} -0.5+0.5i \\ 1 \end{bmatrix}, \; x_2 \neq 0.$$

$$\lambda = 2-i \quad , \quad \mathbf{x} = x_2 \begin{bmatrix} -0.5-0.5i \\ 1 \end{bmatrix}, \; x_2 \neq 0.$$

Before stating Theorem 1, we extend conjugate operation from numbers to vectors and matrices. We say that a vector $\bar{\mathbf{x}}$ is the conjugate of a vector **x** if each entry of $\bar{\mathbf{x}}$ is the conjugate of the corresponding entry of **x**. A similar definition applies to the conjugate of a matrix. The following property of the conjugation operation is easy to prove and we will use it in the proof of Theorem 1:

(2) $$\overline{A\mathbf{x}} = \bar{A}\bar{\mathbf{x}}.$$

Having the result given in Equation (2), we are ready to state and prove Theorem 1.

> **THEOREM 1**   Let A be a square real matrix. If λ is a complex eigenvalue of A with corresponding eigenvector $\mathbf{x}$, then $\overline{\lambda}$ is also an eigenvalue of A with corresponding eigenvector $\overline{\mathbf{x}}$.

Proof: Suppose that $A\mathbf{x} = \lambda\mathbf{x}$, where $\mathbf{x}$ is nonzero. Then, taking conjugates of both sides, we have

(3) $$\overline{A\mathbf{x}} = \overline{\lambda\mathbf{x}}.$$

Now, observe that $\overline{A\mathbf{x}} = \overline{A}\,\overline{\mathbf{x}} = A\overline{\mathbf{x}}$ (the first equality is from Equation (2), the second follows because A is a real matrix). Also note that $\overline{\lambda\mathbf{x}} = \overline{\lambda}\,\overline{\mathbf{x}}$. Using these two results in Equation (3), we obtain:

$$A\overline{\mathbf{x}} = \overline{\lambda}\,\overline{\mathbf{x}}.$$

The equation above shows that $\overline{\lambda}$ is an eigenvalue of A with corresponding eigenvector $\overline{\mathbf{x}}$. ■

Eigenvalues of a symmetric matrix are real

We close this section with a proof of Theorem 1 in Section 3. In particular, let A be a real, symmetric matrix and let λ be an eigenvalue of A with a corresponding eigenvector $\mathbf{x}$. We prove that λ is real by showing that $\lambda = \overline{\lambda}$. In particular:

$$A\mathbf{x} = \lambda\mathbf{x} \qquad \text{(This is given)}$$
$$\overline{\mathbf{x}}^T A\mathbf{x} = \lambda\overline{\mathbf{x}}^T\mathbf{x} \qquad \text{(Multiply both sides by } \overline{\mathbf{x}}^T\text{)}$$
$$\mathbf{x}^T A^T\overline{\mathbf{x}} = \lambda\mathbf{x}^T\overline{\mathbf{x}} \qquad \text{(Take the transpose of both sides)}$$
$$\mathbf{x}^T A\overline{\mathbf{x}} = \lambda\mathbf{x}^T\overline{\mathbf{x}} \qquad \text{(}A\text{ is symmetric)}$$
$$\overline{\mathbf{x}}^T \overline{A}\mathbf{x} = \overline{\lambda}\overline{\mathbf{x}}^T\mathbf{x} \qquad \text{(Take conjugates of both sides)}$$
$$\overline{\mathbf{x}}^T A\mathbf{x} = \overline{\lambda}\overline{\mathbf{x}}^T\mathbf{x} \qquad \text{(}A\text{ is real)}$$

In the sequence of six equalities above, the second and last equality have the same left-hand side. Subtracting the second from the last, we obtain:

(4)
$$0 = \overline{\lambda}\overline{\mathbf{x}}^T\mathbf{x} - \lambda\overline{\mathbf{x}}^T\mathbf{x}$$
$$= (\overline{\lambda} - \lambda)\overline{\mathbf{x}}^T\mathbf{x}$$

Note that $\overline{\mathbf{x}}^T\mathbf{x} = \overline{x}_1 x_1 + \overline{x}_2 x_2 + \cdots + \overline{x}_n x_n = |x_1|^2 + |x_2|^2 + \cdots + |x_n|^2$ and therefore (since $\mathbf{x}$ is an eigenvector) the term $\overline{\mathbf{x}}^T\mathbf{x}$ is positive. Thus, see Equation (4), we have to have $\overline{\lambda} - \lambda = 0$. Equivalently, $\overline{\lambda} = \lambda$ and this means that λ is real.

Exercises 4.5

1. Let $A = \begin{bmatrix} 6 & 8 \\ -1 & 2 \end{bmatrix}$.

   (a) Show that $\mathbf{u_1} = \begin{bmatrix} 2 + 2i \\ -1 \end{bmatrix}$ is an eigenvector for A and determine the value of the associated eigenvalue λ_1. [HINT: Calculate $A\mathbf{u_1}$.]

   (b) Exhibit a second eigenvalue, λ_2, for A and a corresponding eigenvector $\mathbf{u_2}$.

2. Repeat Exercise 1 for $A = \begin{bmatrix} 2 & 4 \\ -2 & -2 \end{bmatrix}$ and $\mathbf{u_1} = \begin{bmatrix} 1 + i \\ -1 \end{bmatrix}$.

In Exercises 3-5, find the eigenvalues and eigenvectors for the given matrix.

3. $\begin{bmatrix} 0 & -1 \\ 1 & 0 \end{bmatrix}$ 4. $\begin{bmatrix} 2 & -1 \\ 5 & -2 \end{bmatrix}$ 5. $\begin{bmatrix} 2 & 1 & -1 \\ 0 & -1 & 1 \\ 0 & -5 & 3 \end{bmatrix}$